Geyelin's Poultry Breeding From A Commercial Point of View
Natural and Artificial Hatching, Rearing and Fattening of Poultry

by George Kennedy Geyelin

with an introduction by Jackson Chambers

This work contains material that was originally published in 1871.

This publication is within the Public Domain.

*This edition is reprinted for educational purposes
and in accordance with all applicable Federal Laws.*

Introduction Copyright 2017 by Jackson Chambers

Self Reliance Books

Get more historic titles on animal and stock breeding, gardening and old fashioned skills by visiting us at:

http://selfreliancebooks.blogspot.com/

Introduction

I am pleased to present yet another title on Poultry.

The work is in the Public Domain and is re-printed here in accordance with Federal Laws.

As with all reprinted books of this age that are intended to perfectly reproduce the original edition, considerable pains and effort had to be undertaken to correct fading and sometimes outright damage to existing proofs of this title. At times, this task is quite monumental, requiring an almost total "rebuilding" of some pages from digital proofs of multiple copies. Despite this, imperfections still sometimes exist in the final proof and may detract from the visual appearance of the text.

I hope you enjoy reading this book as much as I enjoyed making it available to readers again.

Jackson Chambers

PREFACE

TO THE AMERICAN EDITION.

Most of the experiments in keeping poultry on a large scale have either failed entirely or only partially succeeded. What is the cause of failure in such cases? Is it to be found in any inherent difficulty in the system itself, or does it arise from the want of the application of rational principles? Why should not poultry keeping as a business succeed as well in proportion to the number kept as poultry keeping on a small scale? Why may not an individual succeed as well with a thousand inmates of his poultry yard as with fifty or a hundred, provided he gives the proper attention to the individual wants and requirements of each?

This little work attempts to answer these questions, and it appears to do so satisfactorily. It has attracted a great deal of attention, not only in England, where poultry keeping is followed with enthusiasm, but also in this country, where, if it is followed on a smaller scale, it is very generally pursued as a necessary concomitant of every farmyard. No work on the subject has treated of poultry in a commercial point of view, — that is, the keeping of poultry in large numbers, as a business operation, — so fully or so completely as this of Geyelin; and hence it may be said to be the only truly valuable work on the subject.

Poultry will not bear over-crowding any better than other stock. It must have space enough for pure air and ventilation. It is not essential that it should have free and unlimited range. Some breeds of domestic fowl bear confinement well. Some are not disposed to wander far from their headquarters even if they have their liberty, while others are never easy unless they have full freedom to go wherever they please. It is probable that too little attention has been paid to this point in the attempts made to keep them on a large scale, where they must, of course, be subjected to more or less confinement.

Again, so far as we know, most of the attempts made in this direction appear to have had the production of poultry for market as a leading object, while it is a question whether, commercially speaking, it is desirable to grow fowls mainly for meat. There is more profit in eggs as the leading pursuit, leaving the production of meat as secondary or incidental to the primary object.

No doubt the soil of any location has an important influence on the health of poultry. A cold, heavy, undrained locality is quite unsuited to the poultry yard. The soil should be warm, dry, and sunny, one that is good for grass. If plenty of fish offal were accessible, it should furnish a considerable part of the food. As to breed, it might be any of those commonly known as "everlasting layers," of which, perhaps, the Leghorn, sometimes called the White Spanish, or Andalusian, is as desirable as any. To meet with eminent success on a large scale, it would be necessary to resort to artificial hatching and rearing, both of which are perfectly feasible; for no one could afford to rely upon hens for this purpose where the primary object was to obtain the largest number of eggs. Suppose, then, the chickens were hatched from the first of April to the middle of May. At five weeks old they could be turned out and treated according to one of the following systems: —

1st. Enclose grass land in quarter-acre lots, with a small poultry house in each, or a quadruple house in the centre of four lots, with accommodations for fifty hens — never more. Young chickens might do well in somewhat larger numbers through the summer, but it would be safer, as a general rule, to limit the number to fifty.

2d. Build coops of lath or thin boards, about ten feet long, four feet wide, and two feet high, — four feet in length at one end to be a tight house, or coop of boards, with floor and feeding conveniences, water, &c., — the latticed portion to be bottomless. Arrange handles at each end, so that two men could lift and move the whole. Set these coops upon grass ground, and move them their length or width daily, thus affording a fresh grass run. Twelve chickens should do well in each. As soon as they can be distinguished, separate the cocks from the pullets, and *never* allow them together except for breeding purposes afterwards. As soon as the cocks are marketable, sell them, reserving only the best individuals as breeders, with little, if any, regard to consanguinity. Keep an unlimited supply of cracked corn before them until they are large enough to eat it whole, when it may be given them uncracked. This, with grass, is their main diet. Give also some variety with a little animal food. The pullets should begin to lay early in October, when they should have a plenty of fish waste, and lime in some form, in addition to the grain. In twelve months from the time they begin to lay they should produce one hundred and fifty eggs each, and if properly cared for they might do more. As soon as the hens stop laying and begin to moult, kill and sell them. The white Leghorns are always ready for the table.

I do not know that this movable coop has been tried on a large scale; but there seems to be no reason why it should not prove successful. Grass will grow wonderfully under it; and this could be used either for soiling or for hay. Some other conveniences would, of course, be necessary in winter.

A coop of the above-mentioned size would accommodate twelve laying hens; and four of them, with forty-eight hens, would probably do better than the same number in the enclosure plan, and avoid the necessary investment for fences and repairs. Some say poultry in such confinement, when *all* their wants are supplied, will pay better than when running at liberty, either in growth, fat, or eggs; and it is probably true.

Now, if one coop will succeed, or if one enclosure like that described will succeed, what conceivable reason is there why any number should not? We all know that success in anything depends as much upon details as upon plan. Without attention to either, failure is certain. With only one, success can be but partial.

These are only suggestions to those who are interested in the subject. This little treatise is full of suggestions of a practical character, valuable even for those who are keeping poultry only upon a small scale. If I succeed in placing it within the reach of those who have sought in vain to procure a copy, I shall have accomplished my object.

CHARLES L. FLINT,
Sec'y Mass. State Board of Agriculture.

BOSTON, *May 8, 1867.*

PREFACE

TO THE ENGLISH EDITION.

IN writing this essay on poultry breeding, I have endeavored to avoid all technical expressions and the usual verbiage to swell a book. Neither have I touched upon topics which have no immediate concern with the subject matter, but I have confined myself exclusively to giving publicity to such facts as I have proved by *actual experience;* and I firmly believe that this treatise on *poultry breeding*, in a *purely commercial point* of view, is the only one ever published, in this or any other country, from which the public can learn how to enter upon a highly profitable and pleasing undertaking, and this without having to pay the usual heavy penalties of experimenting. I must, however, caution the reader not to be startled by the novel plan of hatching, rearing, and fattening poultry which I advocate, and of which they cannot find corroboration in any other book; and I advise them, before criticising the principles herein put forth, to well weigh all that is stated, when I doubt not that every one will admit that the principles are logical and based upon sound sanitary and scientific laws. To increase the size of this publication, I might, like others, have copied and annexed chapters on diseases and their remedies; on races and their peculiar distinguishing features; on artificial incubation from the ancient Egyptians and Chinese to the present day; on the history of domestic fowls from the Assyrians to 1865; but to publish a voluminous book is not my object.

My desire is to impart to the public in general, in as few words as possible, with the assistance of comprehensive sketches, and even then leaving them ample latitude to engross, certain general rules and matters ascertained by experience on profitable poultry breeding.

Should I be fortunate enough to accomplish this object to the satisfaction of the public, my task will bear its own reward by a rapid development of poultry breeding in England, as well as in other countries, which will add materially to the wealth and comfort of nations.

From the last Trade Returns it appears that upwards of three hundred million of eggs are now imported annually into England! Can anything show more forcibly the immense and profitable field that lies open to English enterprise in poultry breeding?

G. K. G.

BELGRAVE HOUSE,
    ARGYLE SQUARE, W. C.

CONTENTS.

	Page
Considerations on the Necessary Appliances to Successful Poultry Breeding.	13
A Poultry Home. The Open Run.	16
The Glass-covered Run. The Roosting and Laying-Room.	18
The Hatching-Room.	20
Reference to Plan and Perspective Section of the Poultry Home and Vinery.	20
General Rules to be observed in Poultry Breeding.	24
The Laying of Eggs.	26
The Ovarium.	27
Natural Hatching.	28
General Observations on Poultry Food and Drink.	31
The Drink for Poultry.	31
Food for Young Chickens.	32
The Food for the Breeding and Laying Stock.	32
The Food for the Fattening Stock.	33
Preparation of the Fattening Food. Poultry Manure.	34
The Feathers of Fowls. The Moulting of Fowls.	35
Diseases in Poultry.	35
Various Races of Poultry.	36
Killing and Dressing Poultry for the Market.	37
Machinery, Implements and Utensils.	38
Artificial Hatching.	38
The Artificial Hatching-Room.	43
Portable Artificial Hen for Hatching.	45
Artificial Poultry Hens for Rearing Chickens.	48
Reference to Perspective Section of Artificial Hen.	50
Artificial Rearing Home.	51
Reference to Perspective Section of Artificial Rearing Home.	52
Artificial Vermin Nursery.	53
Improved Fattening Pens for Cramming Poultry.	55
Preservation of Eggs.	56
Whitewash. Lime Water.	58
Oxide and Sulphate of Iron. General Plan of Buildings.	59

CONTENTS.

Bird's-Eye View and Section of a Poultry-Breeding Establishment.	60
The Patent Vermin Attraction Trap.	66
ESTIMATE OF REVENUE AND EXPENSES FOR A POULTRY-BREEDING ESTABLISHMENT OF 3000 STOCK FOWLS.	68–70
THE LAWS OF NATURE IN RELATION TO POULTRY KEEPING.	71
Egg Preserving.	71
Patent Pneumatic Self-indicating Air-tight Jars.	75
Packing the Eggs.	76
Why Eggs should be packed with the Small End upwards.	77
Warming Poultry Homes.	77
Our System of selling Poultry.	78
EXTRACTS FROM THE "JOURNAL OF HORTICULTURE AND COTTAGE GARDENER."	83
Home Supply of Poultry and Eggs.	83
Poultry and Egg-preserving Company.	85
Home Supply of Eggs and Poultry.	86
Poultry Keeping from a Commercial Point of View.	88
Poultry and Egg Company.	91
Poultry Keeping from a Commercial Point of View.	93
REPORT OF MR. GEYELIN, MAY 17, 1865.	96
Poultry Breeding.	97
Vegetable Growing or Market Gardening.	98
Poultry Breeding and Vegetable Growing.	99
Estimate of Revenue and Expenses.	99
Proposed Stock.	100
Working.	101
REPORT OF MR. GEYELIN ON THE POULTRY ESTABLISHMENTS IN FRANCE, JULY 10, 1865.	103
The Object of the Voyage.	104
Natural and Artificial Incubation.	107
The Rearing of Poultry.	111
Feeding and Fattening.	113
Killing and Dressing.	116
Utilizing the Waste Products.	118
The System of Selling.	119
The Distinct Breeds.	121
Caponage and Virgin Cocks.	124
Opinions on my System of Poultry Breeding and Rural Economy.	124
Analysis of my Observations.	125

POULTRY BREEDING.

Considerations on the Necessary Appliances to Successful Poultry Breeding.

A UNIVERSAL notion prevails that poultry cannot possibly be bred with profit except on farms, and then only when bred in large quantities. This is a most mistaken idea, as a few heads of poultry will yield proportionately as much profit as any larger numbers. For instance, whereas in large establishments heavy expenses are incurred for buildings, rent, machinery, and labor, these charges do not occur with the amateur breeder who attends on his own poultry personally. It is true that large establishments can buy their cereals cheaper, and grow their own vegetables; but this, again, is compensated with the amateur who obtains a better price for his eggs and poultry, even if used for his own consumption, than the large breeder, who is obliged to sell his produce through a salesman at wholesale prices.

However, to obtain such satisfactory results, it is absolutely necessary to observe certain sanitary laws in the construction of the poultry home, and to see that the dietary scale is conformable to the confined state, a. in fact, providing poultry with an equivalent of such food as they could pick up when in a free state. The poultry home I suggest is applicable alike to amateurs and large breeders, and is intended for the accommodation of one cock and six hens for breeding, or twelve hens for laying, and twenty-four to thirty half-grown chickens; and as the same principle must be carried out, whether in small or large establishments, it follows that where it requires only one home for seven, twelve, or thirty birds, it will require one hundred homes for seven hundred, twelve hundred, or three thousand birds, and so on in proportion to the magnitude of the breeding establishment. This plan has, moreover, the advantage of keeping the races and sexes separate, of affording an easy inspection, and of extending and multiplying the homes gradually with the growth of the establishment, besides facilitating the labor in feeding and hatching, and the sanitary requirements. Another erroneous idea entertained is, that poultry will never thrive well in a confined state; whilst, in fact, they will thrive much better, and be much more productive than when left roaming about in all weathers in search of food, provided the directions given hereafter are implicitly followed: however, it is so far true, that poultry confined in a damp and ill-ventilated place, and having a deficient and ill-adapted diet to their confined state, can never thrive; but whose fault is this? Why, it might as well be said that a person cannot thrive during solitary confinement, when it is well known that prisoners with a regular diet, comfortable cells, and appropriate labor, soon become very sleek and healthful in appearance, and that in proportion there are less disease and fewer deaths in prisons than among the free population who are compelled to seek a precarious living in all kinds of weather, and whose homes are wretched hovels, deficient of all sanitary requirements.

Moreover, farmers have now for a number of years carried on successfully the rearing and fattening of cattle in confined spaces (which are called stall-fed cattle), and which system, although nominally more expensive, is yet far more profitable than the ordinary rearing of cattle; and why should the same system not be extended to poultry?

In general, the management of poultry is considered of too little importance, and is left pretty well to chance: it is true that of late years the poultry exhibitions have created a taste for poultry breeding; but this is confined solely to amateurs and what may be called fancy poultry breeding. Yet, amongst all domestic animals, the fowl is in proportion to its cost or keep the most profitable and useful; and hereafter I will prove by figures obtained by actual experience that poultry can be reared and sold at the rate of four pence per pound, and leave a handsome profit. Now, such results — particularly when butchers' meat is at ten pence and one shilling per pound, and moreover daily rising in price on account of the increase of population and the decrease of pasturage — ought to prove a sufficient stimulant to the public at large to give a little more attention and consideration to an increased production of such valuable animal food, which, by proper management, would, within a very few years, become as much the food of the poor as it forms now a delicacy for the rich only.

A Poultry Home

Ought to consist of four separate compartments, exclusive of the glass-covered passage which runs the whole length of the building, to facilitate the service at all times and weather.

1st. A roosting and laying compartment.

2d. A glass-covered run, in which they can be confined in wet weather.

3d. A hatching-room.

4th. An open run.

The Open Run.

Starting with the well-ascertained fact that poultry cannot possibly thrive or be kept in good health on damp ground, it becomes necessary, where there is no surface gravel land, to make an artificial dry run: this is best accomplished with concrete, which, besides being cheaper than flag-stones or bricks, does not absorb the moisture, and is much warmer to the feet.

This run should be formed slightly concave, as shown by Fig. 9, and have an incline towards its end, where the rain-water can be discharged into a drain communicating with the duck pond.

During summer a few inches deep of gravel, and in winter about nine inches deep of horse manure, ought to cover the floor of this run, which will afford the fowls ample exercise by scratching and running. The gravel and horse manure should be turned over at least once a week, and renewed whenever they become too much impregnated with the fowls' droppings. This will prevent the diseases which a tainted ground gives rise to among poultry. The sides and top should be formed of galvanized iron-wire netting of about one and a half inch meshes for full-grown poultry, and one inch meshes for chickens. In fine weather the food should be thrown broadcast on this run; but in wet weather the poultry ought to be fed from feeding-vessels placed in the roosting-room, and near to the door of the covered run.

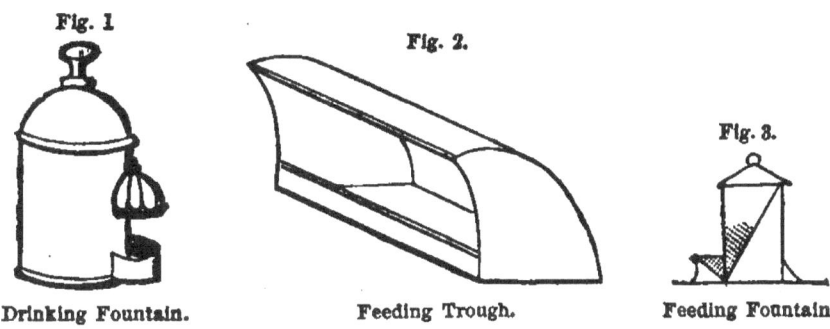

Fig. 1 Drinking Fountain. Fig. 2. Feeding Trough. Fig. 3. Feeding Fountain.

The above fountains are constructed on the principle that when placed with their opening towards, and about one foot from, the wall, the poultry will be unable to scratch any dirt in, nor can the droppings from the perches fall in.

Some persons advise boarding the sides of the run to the height of two feet, to prevent the cocks of the adjacent run from fighting together; this, however, in so narrow a run, would partially exclude the sun, which is not desirable; besides, cocks after a few days' acquaintance become very neighborly.

The Glass-covered Run.

The floor of this compartment should be composed of at least six inches deep of finely-sifted gritty stuff, such as road dust, ashes, and sand, and on this it would be well to sprinkle occasionally some flour of sulphur, which would prevent vermin breeding on the fowls. It is the universal belief that fowls powder themselves to get rid of vermin on their bodies; but such is not exactly the case. The fact is, fowls, like all other feathered tribes, perspire freely, particularly so during night time: this perspiration clogs their feathers; consequently they perform their ablution in gritty dust on the same principle that human beings do in water, to get rid of the dried perspiration and to expand their feathers. The same can be seen performed by the birds of the air, which, on a dry, hot summer day, make their ablutions in the dust of the roads.

In this run the fowls ought to be kept close during wet weather, as wet feathers are most injurious to their health, particularly when in a confined state.

The Roosting and Laying Room

Should be kept scrupulously clean, swept out daily, and occasionally thoroughly whitewashed, the floor slightly sanded over daily.

The nests, in a sanitary point of view, ought to be made of earthen ware, partly filled with fine sand or cocoa-nut refuse, and slightly sprinkled over with flour of sulphur. The roosting perches should be formed of hot-water pipes, as they are of the utmost importance to keep the poultry warm during the cold nights, and cool during hot nights, and which will induce a continuous laying of eggs during a time when they are most scarce either for hatching or consumption.

Most persons must have observed that even the heaviest fowls will seek to perch nearest to the ceiling, and that when roosting their feathers are ruffled or open. This is easily explained by all persons conversant with the aerostatic laws; namely, that heated air being lighter than cold air it will ascend; consequently the warmest place in a room will be nearest to the ceiling; therefore fowls open their feathers when roosting to admit the warm, ascending air. Another important point in the construction of this room is the creation of a perfect ventilation without causing any draught. Different gases, varying in their specific gravity, are formed in this room, namely, carbonic acid, which is a heavy gas and hangs near the floor, the ammoniacal gas from the excrements of fowls, and carburetted hydrogen gas from the exhalation of the fowls, both of which gases are light, and consequently rise to the ceiling. It becomes, therefore, necessary to adopt a principle of ventilation by which both the heavy and light gases can be got rid of without causing a draught, which would be prejudicial to the health of the fowls. This is accomplished by two pieces of perforated zinc, one opposite to the other, near the floor, and the same near the ceiling, and at least twelve inches above the roosting perches.

The Hatching-Room,

In my plan, is situated immediately above the roosting-room, and composed of two compartments — the one in which the hen sits, the other where she has a supply of gritty dust to perform her ablutions.

Reference to Plan and Perspective Section of the Poultry Home and Vinery.

A is a glass-covered passage running the whole length of the building, and from which communication is obtained by means of doors to all the compartments of the homes on either side. This passage ought to be about six feet wide and eight feet high to the rise of the roof.

a is a flue formed of bricks and covered with paving-tiles, with ventilation at certain distances. This flue runs the whole length of the building, and ought to be about nine inches wide and fifteen inches deep: it serves for warming the building by means of hot air, steam, or hot-water pipes, and the admission of heat is regulated by means of the ventilators.

The floor should be formed of concrete, the sides of whitewashed boards, and the roof of glass with movable frames at certain distances to allow of ventilation. This passage can be turned to a profitable account by being used as a vinery or conservatory without extra cost.

B is the roosting-room, about three feet square and six feet high. The floor should be made of concrete,

A COMMERCIAL POINT OF VIEW.

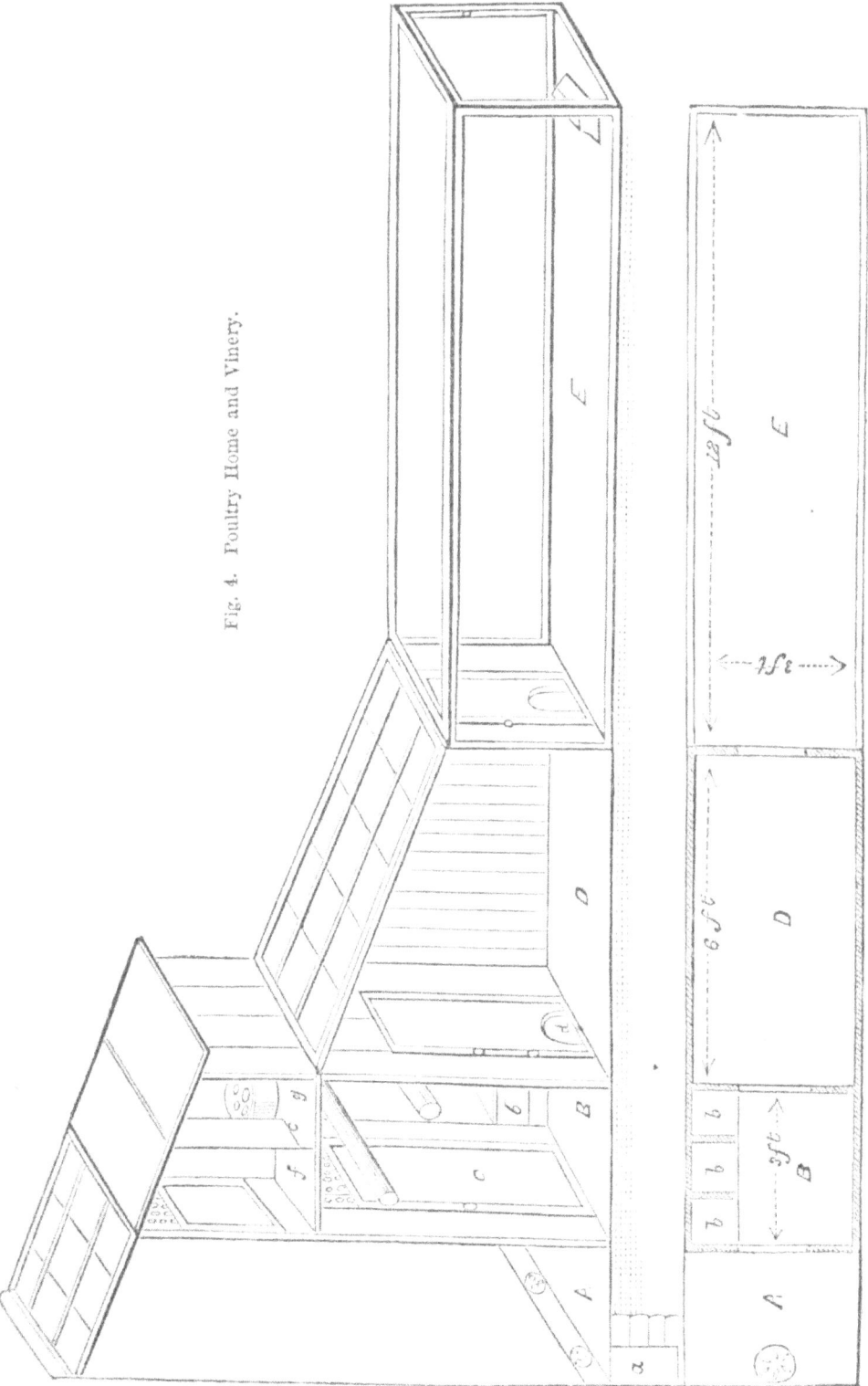

Fig. 4. Poultry Home and Vinery.

the sides and ceiling of whitewashed boards; the partitions of the nest should also be made of whitewashed boards, and the nest of earthen ware; but the top board covering the nest should project a few inches to prevent the droppings falling in.

c is a door communicating with the passage, and *d* with the covered run. In this door an opening ought to be made provided with a glazed slide for the egress and ingress of the fowls. In this compartment fowls should be fed in wet weather, and the drinking fountain ought also to be placed here. The perches of cast-iron pipes should be about three inches in diameter, and placed respectively three and four feet from the floor.

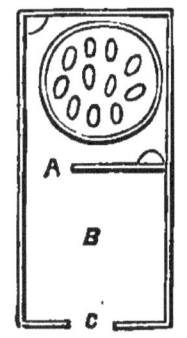

Fig. 5. Hatching-Room.   Fig. 6.

C, the hatching-room, is composed of two compartments — one for the gritty dust, and the other for the nest, which should be of earthen ware, the same as for the laying nests. The floor, sides, and ceiling are of whitewashed boards. The compartments are eighteen inches square by two feet high, the door glazed, and with perforated zinc above for ventilation; the roof covered with asphalted felt.

D, the glass-covered run, should not be less than six feet long, three feet wide, four feet high to the rise, and six feet to the apex on top of the glass frame, which ought to be movable to admit of ventilation. The sides should be formed of whitewashed boards. A perch can also be fixed with advantage in this compartment.

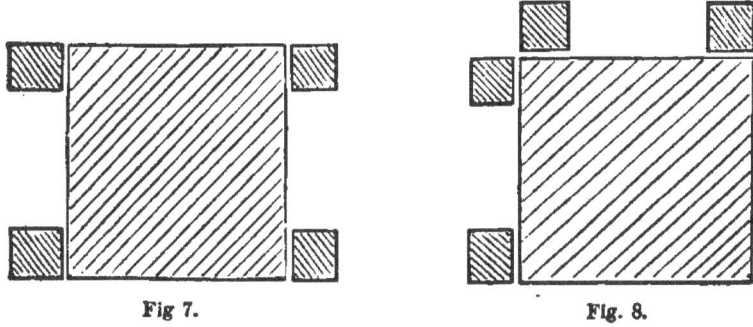

Fig 7.   Fig. 8.

E, the open run, should be not less than twelve feet long, three feet wide, and three feet high. The plan I recommend for the construction of open runs consists of separate wooden frames six feet by three feet (see Fig. 6), on which the wire netting is fixed, and grooved

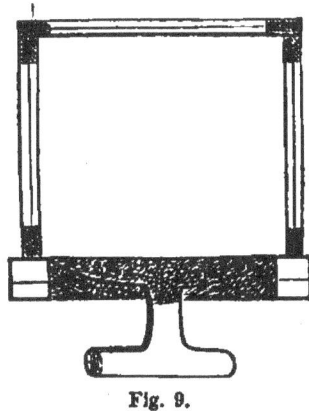

Fig. 9.

uprights, in which these frames are slid. (See Figs. 7, 8.) The frames forming the top can be joined together by hinges. By adopting this plan, the whole run can be removed in a few minutes, or any part can be taken away for repair without interfering with the others, or some spare frames might even be kept in stock to replace those that want repairing.

General Rules to be observed in Poultry Breeding.

The Breeding Stock.

1st. The hens selected to breed from should be kept apart from the cock until they are at least twelve months old; and the cock should not be less than eighteen months old before he is put with hens, as a too early call on nature degenerates the breed.

2d. Whatever races are selected, they should be the most perfect specimens that can be obtained, as the first outlay will repay itself.

3d. That the distinct races be kept strictly separate except where it is intended to obtain a cross breed; and for this the finest specimens of both races and sexes should be selected.

4th. Not more than six hens should be allotted to a cock.

5th. After the third breeding year it is advisable either to sell the stock or to fatten them for the market, as they become less fecund, and their progeny are apt to degenerate.

6th. The eggs should be collected at least three times a day, as in a fecundated egg, when set upon for a few hours, the germ very soon gets developed, and the egg is afterwards unfit for hatching.

7th. The stock must be fed regularly at sunrise and the afternoon an hour before going to roost.

The Laying Stock.

1st. When it is intended to sell the eggs for consumption it is advisable to pen hens up without a cock to prevent the eggs being fecundated, as they will then keep fresh much longer; this system of keeping hens by themselves has another great advantage, as they will lay a great many more eggs during the year.

2d. About twelve to eighteen hens can be kept together in a home, as shown by Fig. 4.

3d. The eggs should be collected twice a day.

4th. For feeding, the same rule applies as above; and the reason for selecting sunrise and afternoon for feeding time is, that it is before and after the laying time, during which the hens on their nest would get no food.

The Chickens.

1st. From the time they are hatched to the time when they begin to roost, not more than twelve chickens ought to be kept in one compartment, as they will huddle together, and the weak ones either get crushed or suffocated.

2d. The place where the young chickens retire to ought to have a dry floor, and be kept scrupulously clean; and as the floor is the coldest part of a room, their roosting-box ought not to be more than twelve inches high, and to be slanting, which will keep the warm air in the roost. (See Fig. 14.)

3d. As soon as they begin to roost on perches, they can be removed to a poultry-home, say about thirty to each home.

4th. When the cockerels can be distinguished from the pullets, they should be penned up separate. From this stock the breeding and laying stock will be selected to replace old ones.

5th. The feeding of chickens ought to take place not less than three times a day, and be of a liberal kind, with plenty of finely-chopped green vegetables, and an occasional supply from the vermin nursery, but no meat should be given.

6th. Occasionally a little flour of sulphur and oxide of iron mixed with their food will keep them in good health, also sulphate of iron and lime water in their drink. The same is applicable for all kinds of poultry.

The Laying of Eggs

Takes place in the morning during the summer months, and gradually later in the day as the winter approaches, until moulting time arrives, when the hens cease laying till they have their new feathers, which takes about two months.

Although a hen can only lay a determined number of eggs during her lifetime, yet her laying may be stimulated by an appropriate diet (see Food), as also by a genial temperature kept in the poultry-home.

It has been satisfactorily proved that under such circumstances a hen will lay at least thirty eggs more during the winter months, a time when they are most valuable both for artificial hatching and consumption; and taking an establishment with two thousand laying and one thousand breeding hens, the extra profit will be as follows: Three thousand hens at thirty extra eggs equal ninety thousand at 15s. per hundred, £675, to be ascribed solely to a warm temperature and appropriate diet; but this is not the only advantage derived from a genial temperature during the winter months; it may save, perhaps, hundreds of pounds in the loss of poultry from diseases caused by exposure to damp and colds.

As the laying can be forced by artificial means, so can it also be retarded; and when it is intended to keep some hens for laying during the time that others are moulting, which generally begins in September, it is only necessary to pull out the feathers of such hens, and thus produce an artificial moulting about two months sooner, say early in July, when they will cease laying until their feathers have grown again.

The Ovarium.

It has been ascertained that the ovarium of a fowl is composed of six hundred ovulas or eggs; therefore a hen, during the whole of her life, cannot possibly lay more eggs than six hundred, which in a natural course are distributed over nine years in the following proportion: —

First year after birth,		15 to 20
Second " "		100 " 120
Third " "		120 " 135
Fourth " "		100 " 115
Fifth " "		60 " 80

Sixth year after birth, ... 50 to 60
Seventh   "   "   . . . . . 35 " 40
Eighth   "   "   . . . . . 15 " 20
Ninth   "   "   . . . . . 1 " 10

It follows that it would not be profitable to keep hens after their fourth year, as their produce would not pay for their keep, except when they are of a valuable or scarce breed.

Natural Hatching.

The hens of all kinds of gallinaceous fowls sit for twenty-one days; ducks of the usual kind, such as Aylesbury, Rouen, and others, twenty-eight days; Muscovy ducks, thirty to thirty-five days; geese, thirty to thirty-five days; Guinea fowls, twenty-eight to thirty days; turkeys, twenty-eight days; pea hens, twenty-eight to thirty days. With a view of obtaining more eggs in a given time from a fowl, many writers suggest to prevent the hen from sitting by cooping her up in a dark place on a low diet. Nothing can be more cruel than to force nature without giving that necessary rest which overwork requires. Already the domesticated fowls lay many more eggs than wild ones between their hatchings, and by a judicious housing and feeding, can be made to lay still more; but then it is absolutely necessary to allow her to recruit her strength by a rest of twenty-one days on her nest, and a liberal poultaceous diet, as the laying of eggs, and more particularly of large ones, is attended with considerable pain, as is evidenced by the difference of sound hens utter before and after their laying, and also from their uneasiness whilst on their nest. Besides, domesticated fowls are naturally of a sociable disposition, and to separate a hen from her companions, and to keep her on a low diet when she requires rest and nourishing food to recruit her strength after she has become exhausted from the pain of laying and the drain on her constitution by the rapid formation of eggs, is the height of cruelty, and would surely not be practised were breeders aware of the injury they do the health of their hens. I do not say that hens should be permitted to rear their brood, as that would be waste of time, and most hard work to a kind mother, who will but ill feed herself to provide as much as possible for her young; she has rest neither day nor night, as she is compelled to remain in an unnatural position to cover her young ones. The rearing can be performed with greater success by artificial mothers, as will be explained hereafter; but what I do advise those persons who have a regard for the health of their fowls, and their own interest into the bargain, is to allow Nature her own way by giving a hen her twenty-one days' rest, and the while a quiet place and nourishing poultaceous food; after which time she can be returned to her own home, when in a few days she will recommence laying.

When a hen wants to sit she utters a peculiar cluck, ruffles her feathers, and wanders about, searches dark corners, and is evidently ill at ease; she is feverishly hot, and resolutely takes to a nest in which there are eggs, whether of her own production or not matters little to her; at this time a hen will allow herself to be separated from her companions, and placed in solitary confinement, without fretting, provided she has a nest and eggs to sit upon. It is not advisable to allow a hen to hatch in her ordinary home, and amidst her companions, who are fond of usurping the nest, and laying fresh eggs in it.

A warm moisture being necessary to the hatching of strong and healthy chickens, as evidenced by wild birds and hens that sometimes unobserved will hatch a brood under a hedge in the fields, I recommend the sitting nest to be made of earthen ware, the same as for laying, with this difference, that a fresh-cut piece of turf should be placed on the sand, and on which the eggs are put; the heat of the hen will soon generate steam, but whenever the turf gets too dry, some water may be poured on the sand underneath.

The number of eggs to be placed under a hen must necessarily depend on her size. A Dorking, Cochin China, or Bramah Pootra, or other large breed, can with every certainty hatch at least fifteen eggs; and as regards the selection of eggs, all I can advise is to select fresh and good-sized ones.

Some persons pretend to be able to tell whether an egg is fecundated, and whether it will produce a male or female bird; but these assertions have as yet not been satisfactorily proved.

Fig. 10.

General Observations on Poultry Food and Drink.

When poultry is kept in a confined state its food must be appropriate. A fowl kept in a free state on a farm can with advantage be fed all the year round with barley or oats only, as she will supplement her meals with animal and vegetable matters of her own finding; therefore an equivalent should be given to penned-up poultry; but again, as they have not so much bodily exercise as when in a free state, their digestive powers are weakened, consequently they are subject to inflammation of the bowels when fed on whole grain only. After this explanation, my readers will understand the reason why I advocate all grains to be ground, and the meat and vegetables to be minced; but apart from the sanitary consideration, it becomes an important economical fact in a large breeding establishment, as it is well known that poultaceous food made of pounded grain, and which calls little on the digestive organs, has far greater feeding and fattening qualities than the whole grains.

There is another point connected with the feeding to which I wish to allude. The diet should be varied almost daily, but green vegetables finely minced ought to form part of every meal, and occasionally some oxide of iron, and at other times flour of sulphur, mixed with their food will greatly tend to keep poultry in good health.

The Drink for Poultry.

The water should be changed daily, and occasionally clear lime-water, and at other times sulphate of iron mixed with it.

Food for Young Chickens.

Indian and barley meal, boiled rice, mashed potatoes, bread crumbs, &c., steeped in milk and water; any of the above, separate or mixed together, will do well.

Finely-chopped green vegetables daily, and occasionally hard-boiled eggs chopped fine, with a supply from the vermin nursery. The water should be supplied between two saucers (see fig. 11) to prevent wetting themselves,

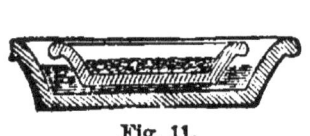

Fig. 11.

Fig. 12.

or to scrape the food out. The inner saucers can be partly filled with coarse sand.

Clean water and a plentiful supply of food given about four times a day, and with the comfort of the artificial mothers, chickens will keep in better condition than when left to roam in search of food with a hen.

The Food for the Breeding and Laying Stock

Can be composed of a mixture of the various cereals, coarsely ground, and made into a stiff paste. This food should be put in the feeding fountains, where it cannot be wasted or dirtied. Occasionally, in fine weather, whole grain can be thrown broadcast in the open run.

Finely-chopped vegetables, such as the waste of the kitchen garden, mangold-wurzel, swedes, &c., in a green state, mashed boiled potatoes, and rice; minced boiled meat mixed into a paste with the liquor from the meat, and seasoned with salt, pepper, finely powdered oyster shells, or a little chalk, forms a genial condiment. The broken victuals from hotels, large establishments, &c., can also be used with great advantage for the food of poultry.

Powdered charcoal, oxide of iron, and flour of sulphur, mixed alternately at certain intervals with their food, will keep them in perfect health.

The Food for the Fattening Stock.

As they are still more closely confined, they require a poultaceous diet of a highly fattening nature and of easy digestion. When once poultry is penned up for fattening the diet ought not to be varied.

There are three different ways adopted in feeding poultry for fattening: —

1st. A free feeding, consisting of supplying a fowl with food and water *ad libitum*. This takes much longer time, is more expensive, and less satisfactory in the flesh.

2d. Forced dry feeding, which consists of cramming the fowl with pills of poultaceous food twice or three times a day, and giving water *ad libitum*.

3d. Forced liquid feeding consists of reducing the poultaceous food to a liquid state with milk and water, then to pour it down the fowl's gullet, by means of a funnel, three times a day, and not to supply them with any water.

Preparation of the Fattening Food.

Barley meal, or mixed in equal quantity with Indian meal, made into a stiff paste with milk and water, and seasoned with bay salt.

This paste is then either made liquid, for liquid feeding, or into pills, which should be dipped into milk and water before they are given, so as to facilitate the swallowing.

Experiments have proved that the seasoning poultry food with bay salt produces the following advantages: —

1. To render the fattening of shorter duration.
2. To produce, with the same quantity of food, more flesh and fat.
3. To give the flesh greater firmness and flavor, and to the fat more compactness and a finer grain.

Molasses or sugar mixed with the meal has also good fattening qualities. The duration of fattening must much depend on the condition, age, and health of the fowl, and in this, the same as in administering the food, actual experience is the best teacher, as no rules can well be laid down.

Poultry Manure or Guano.

With the ordinary way of breeding poultry, their valuable manure is lost, and we actually send ships to the Pacific, and all over the world, to fetch those very droppings of fowls which we despise to collect at home. Yet, on a large breeding establishment, the collection of this manure, so much sought by florists, will yield a considerable extra profit, which can safely be calculated at the rate of fifty pounds per thousand fowls annually; but as vegetable growing, the refuse of which is good food for poultry, is almost a necessary adjunct to a large breeding establishment, this manure would be still more valuable to the proprietor on his own land.

The Feathers of Fowls

Are another source of profit in large establishments, where they can be sorted and dried, as they will then fetch a much higher price, and may be computed at ten pounds per thousand heads.

The Moulting of Fowls

Is classed by many writers on poultry under the head of diseases, which it is not; but is only a natural process with most animals in changing their summer coat for a winter one: nevertheless, it is a great drain on their constitution, and fowls, during moulting time, ought to be kept warm, and liberally dieted with warm and stimulating food, such as boiled oatmeal seasoned with salt and pepper, chopped onions, mashed potatoes, and occasionally bread crumbs soaked in strong ale or weak gin. Oxide of iron, lime water, and sulphate of iron can also be given with advantage. This diet will accelerate the moulting, and produce a speedier resumption of laying.

Diseases in Poultry.

Most books on poultry contain a more or less lengthy description of the various diseases fowls are subject to, and prescribe certain remedies; all of which help to swell a book, but are perfectly useless for all practical purposes. We might as well try to doctor ourselves for diseases of which we know nothing.

The diseases in fowls may chiefly be ascribed to our variable climate, to dampness and cold, to injudicious feeding, and to ill-ventilated roosting-places.

A diseased fowl, as will have been observed by many, is never kindly treated by its healthy companions, and, in my opinion, the best and most economical cure for a diseased fowl is to kill her before she gets too far gone, and whilst yet fit for the market; and if not fit for the market, she will, when hacked up, make good food for the pigs.

I acknowledge myself ignorant of the diseases in fowls, consequently of their proper treatment; and as I have no wish to teach the public that which I do not understand myself by simply copying from other books, I shall only state that, with judicious feeding and housing, according to my plan, there ought not to be one diseased fowl in a thousand.

Various Races of Poultry.

On this subject I would refer the reader for the desired information to some special publication, as it does not exactly enter into the considerations of poultry breeding in a commercial point of view.

All that is necessary to know of the different races is to be able to distinguish those that are the best layers, the best setters, and the best table fowls, and never mind about the particular points or feathers, the distinguishing characteristics of a fine breed.

Now, where eggs are the sole object, some small breeds

lay larger and more eggs than larger fowls; for this, Hamburgh, Spanish, and some cross breeds may be kept with advantage. As for fowls that will give credit to the breeder for their weight after being fattened, Dorkings, Bramah Pootras, and Cochin Chinas, and their crosses should be selected.

Killing and Dressing Poultry for the Market.

Almost every locality has its own system, but I may advert to a few facts on this subject: Poultry, when bled to death, is much whiter in the flesh. I should advise the following plan as the very best, causing instant death without pain or disfigurement:—

Open the beak of the fowl, then with a pointed and narrow knife make an incision at the back of the roof, which will divide the vertebræ, and cause immediate death: after which hang the fowl up by the legs till the bleeding ceases; then rinse the beak out with vinegar and water. Fowls killed in this manner keep longer and do not present the unsightly external marks as those killed by the ordinary system of wringing the neck. When the entrails are drawn immediately after death, and the fowl stuffed, as they do in France, with paper shavings or short cocoa-nut fibres to preserve their shape, they will keep much longer fresh. Some breeders cram their Poultry before killing to make them appear heavy; this is a most injudicious plan, as the undigested food soon enters into fermentation, and putrefaction takes place, as is evidenced by the quantity of greenish putrid-looking fowls that are seen in the markets.

Machinery, Implements, and Utensils.

Without desiring to recommend any particular plan for the saving of labor, it is yet desirable to state that in any establishment of magnitude the expense of labor forms a prominent item, and that it will therefore be to the interest of the proprietor to invest a certain capital in the purchase of such machines and utensils as will not only economize labor, but also perform the work much better than it could be done by manual labor.

The principal machines required are a grinding mill for the grain, a pug mill for mixing the poultaceous food, a mincing machine for the meat and vegetables, a potato-mashing machine with wooden rollers, a sifting machine for sand and vegetables, a weighing machine, scales, and sundry smaller machines.

Also a steam-boiling apparatus, a heating apparatus, and in fact such appliances as will not only economize labor but also materials, and particularly fuel.

The manual labor itself ought to be subdivided in such a manner that each person has a particular branch to attend to, by which every one will very soon become so expert in the special duty, that the work will be performed much better and in less than half the time.

Artificial Hatching.

Let it be well understood from the onset that I do not advocate artificial hatching and rearing in exclusion of the natural method, but solely as an absolutely necessary accessory in any large breeding establishment. Take, for instance, one thousand breeding fowls; they will lay about one hundred and fifty thousand eggs per annum under ordinary circumstances. Now, supposing a fowl to sit twice in the course of the year, she could, therefore, not rear, allowing for casualties, more than twenty chickens: this would give only twenty thousand chickens per annum; whereas, with the assistance of artificial means, the remaining one hundred and thirty thousand eggs could also be hatched, and in lieu of twenty thousand there could be produced at least one hundred and thirty thousand chickens, allowing also for casualties. What a result from science applied to practical purposes!

Sceptics will of course say it looks very well on paper, but it will never do — it has been tried before and failed. Now, for such reasoning there are endless facts that have forced themselves upon public consideration under similar circumstances; to my own recollection I have heard manufacturers say that they should never give up hand-looms for power-looms, that the goods turned out did not come up to hand-woven: I have seen those who refused to follow the current of improvements swept away from the list of once notabilities.

Up to this very day many object to gas, and will not allow it to be a great improvement on our old oil-lamps; yet were gas ceased to be manufactured to-morrow, what would be the general feeling? For railways and steamboats to cease running, and to have to revert to our old stage-coaches and sailing-ships, would be not only intolerable, but perfectly impossible.

I might adduce hundreds more parallels, with a view

to prove to sceptics that improvements are not only absolutely necessary in all that relates to our comfort, particularly towards an increase in our food, but also that they are perfectly unavoidable, as many farmers who at first resisted the improvements in farming by drainage, machinery, and applied chemistry, have found to their cost. Therefore, in adopting the expression of artificial means, as more readily understood, I do not mean to convey that it is an entirely distinct mode of breeding poultry, but solely an addition to the mode already adopted, and without which poultry breeding can neither be carried on to a large extent nor with great profit.

My intention at first was to divide this treatise in two parts — the first to rearing poultry in a natural way, the second by artificial means — with a view to please those of my readers who object to any artificial means; but in vain have I endeavored to draw a line where natural means end and artificial means begin. The fact is, the domesticated fowl's life is as much artificial as our own mode of living. In truth, with the progress of civilization we insensibly and gradually create for ourselves artificial wants, which by degrees become absolute necessaries, and amongst a thousand others I may mention tea, coffee, potatoes, sugar, tobacco, &c.; and for the cheap production of such necessaries we create artificial labor (machines), steam-power, and artificial manure. Yet with all this evidence of steady progress and improvements before them, and in the current of which they are drawn and carried onwards without knowing it, there are numbers of even well-informed persons who ridicule anything new as preposterous — a sure failure, not wanted; the old thing is the best after all; and yet these very persons are a living evidence against their assertions. True they will never be found among the pioneers of progress, which, if their shallow minds could possibly arrest or hinder, they would too gladly do; but they can no more help themselves being dragged in the wake of progress than they can stop the revolutions of our earth or the tides of the sea. From such persons we should never have had steam-power, railways, telegraphs, machinery, &c., to economize and multiply labor, to annihilate space and time; and yet these persons share in the benefit such improvements have created with the greatest composure, taking them as *faits accomplis*, never giving a moment's thought that but a short time ago they were what they choose to call new-fangled things; they forget that the very clothes they wear, the food they eat, and the beverages they drink are mostly obtained in their superior and cheap form by artificial means; that, in fact, chemical and mechanical results are combinations of artificial means. For the raw materials we must, of course, depend on Nature; but even those we can in some measure improve by art.

Therefore, when I speak of breeding poultry by artificial means, I do not wish to convey that eggs (the raw material) can be produced without a hen; but, when we have eggs, to produce chickens, and from chickens fowls, by a wise appliance of such laws and combinations as science teaches us, as superior to brute care as much as artificial labor by machinery is superior to manual labor, as hot-house-grown fruits and flowers excel those grown in the open air, and as stall-fed cattle are superior to those from the pasturage.

There is nothing absolutely new under the sun; even hatching chickens by artificial means has been carried on in Egypt, China, and other Eastern countries from the remotest ages to the present day: yet in England it has hitherto proved a failure in a commercial point of view. It is true that in Egypt, where they hatch many millions of poultry annually, artificial hatching is a trade of itself, carried on by many hundred proprietors of ovens; and their successful hatching will be apparent when it is stated that they sell one hundred newly-hatched chickens for about three shillings, or that they will return sixty chickens for every hundred eggs intrusted to them for hatching, free of charge. It is also true that the climate and soil in Egypt are more favorable than in England to the rearing of poultry; but then why should we not appeal to science to assist us in overcoming the drawbacks of our soil and climate? No doubt we shall never be able to produce poultry as cheap as in Egypt, where climate, soil, labor, and cost of land are eminently favorable to a cheap production; but in compensation we can get far higher and in proportion more remunerative prices for our poultry, their feathers and manure. It is an acknowledged fact that the artificial hatching of eggs in England, although carried out on principles not in strict harmony with natural incubation, has yet proved far more successful than the artificial rearing of chickens. This, of course, is ascribable solely to the improvident way chickens are treated before they have their natural protection, their feathers, in a climate where the sudden changes in the temperature of the atmosphere, and the almost everlasting humidity of the soil, act prejudicially on young animal life; but surely these are difficulties which can easily be overcome? Do we not produce in England, by artificial means, as splendid tropical fruits as any tropical climate can produce? And why not surpass Egypt in rearing poultry — if not in cost, at least in quality and in scientific feeding and fattening, for which far more remunerative prices are obtained? Well, all this can now be accomplished in England by any person who will follow my plan of hatching, rearing, and fattening poultry by artificial means.

This plan must necessarily be modified according to the importance of the breeding establishment, and the number of eggs to be hatched daily from one to a thousand; but the main principles of a successful artificial breeding of poultry will under any circumstances remain the same.

The Artificial Hatching-Room.

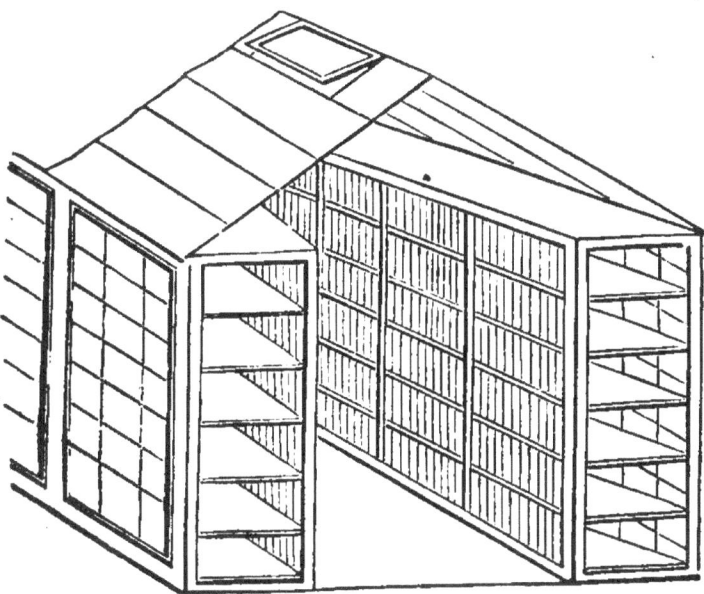

Fig. 13. Perspective Section.

The floor of this room should be of concrete, the sides of movable glazed frames, and the roof of boards covered with asphalted felt, slates, or zinc. The interior ought to be fitted along the sides with movable shelves, which can be drawn out for cleaning; these shelves will be divided into separate compartments three feet long, two feet wide, and one foot high; the sides should be made of galvanized iron wire; so also the front, which forms a door. In each compartment ought to be a frame lined underneath with long fleece, the same as in the portable artificial mothers. In these compartments the chickens are placed from their birth up to a week or ten days old, after which they are put under the care of an artificial movable hen, in small establishments, or in the rearing home in large establishments. (See fig. 17.) These compartments ought to be covered with felt carpet, which must, however, be kept well cleaned, and occasionally dipped in boiling water.

Fig. 14.

The best way to supply food and water to so young chickens is by means of two saucers, one within the other, between which the food or water is put. This will prevent their wetting themselves or scratching the food about. (See fig. 11.)

This hatching-room will require no heating apparatus, as the heat from the hatching apparatus, which is kept in the middle of this room, will keep the temperature sufficiently high during winter.

Near the ridge of the roof ventilating frames should be fixed, and near the floor one or two sliding doors should be provided to allow of the admission of cold air.

Chickens hatched in a dry atmosphere will never be so strong and healthy as those hatched in a moist temperature, as is evidenced by the difference in the appearance of a brood hatched in a loft and one hatched in a field; and as a moist temperature is highly desirable it should be provided for in artificial hatching.

PORTABLE ARTIFICIAL HEN FOR HATCHING.

The apparatus represented by fig. 15, although only

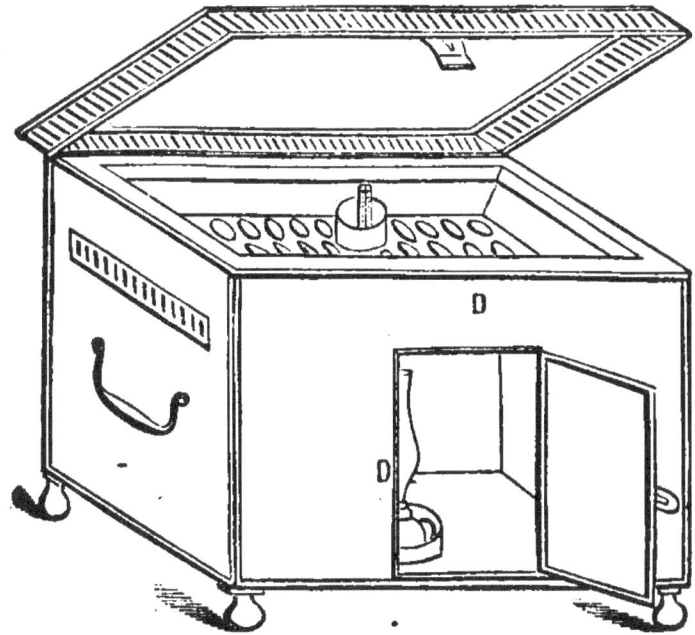

Fig. 15. Perspective Elevation of Artificial Hen for Hatching.

calculated to hatch one egg per day, combines the same advantages as one capable of hatching a thousand eggs per day, and will answer all the requirements of an amateur breeder; besides, it is so portable and convenient in its construction that it can be placed in a bed-room, which, while hatching, it will keep warm day and night, at an equal temperature, and the light from the gas or lamp will serve as a night-light.

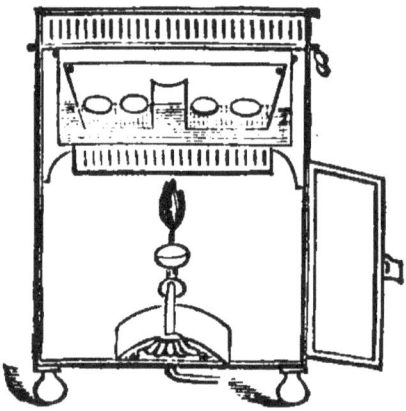

Fig. 16. Transverse Section.

From the above perspective elevation and section it will be seen that the hatching apparatus consists of separate parts.

1. A glass-covered box.
2. A water-tank.
3. A floating vessel.
4. A gas or oil lamp.

The glass-covered box is made of japanned tin; it has a glass door through which the light can be seen; the bottom of this box is perforated in the centre for the admission of air to the lamp, and the other part is car-

A COMMERCIAL POINT OF VIEW.

peted to receive the chickens as they leave their shells. About twelve inches from the bottom are four brackets to receive the water tank; the lid has a perforated border for the escape of the vitiated air and steam from the water. The sides are provided with handles for carrying the box from one place to another, and it stands on four knobs to allow a free passage of air underneath.

The water-tank is made of tin, and a little smaller than the box, so as to allow about half an inch free passage of air all round.

The floating vessel is also made of tin, and is a trifle smaller than the water-tank, so as to allow of its floating in it. The centre of this vessel has an oval opening, in which a registering thermometer is kept to show at all times the temperature of the water. The bottom of this vessel is covered about one inch deep with silver sand, on which the eggs are placed. By means of the central opening, and that between the tank, the temperature is kept in a constant moist state. The lamp can be for oil or gas, but gas is certainly preferable.

The management of the apparatus is so simple that it can be attended to by a child, and only a very few directions will be necessary:—

1. Fill the tank with hot water till the floating vessel reaches the top level, then see that the water has a temperature of about one hundred and twelve degrees, after which light the lamp, and should the heat of the water increase, reduce the flame; but if the temperature rises or decreases but slowly, it can be regulated by admitting more or less air through the door of the box.

2. The principal point, however, is, that the tempera-

ture on the sand should not vary much from one hundred and five degrees, and it will be found that with water-heat of one hundred and twelve degrees, the sand will be one hundred and five, and on the eggs ninety-eight degrees. For beginners, however, it is always best to put the apparatus in action a day or two before placing eggs in it.

3. Turn the eggs once or twice a day, and keep the water replenished as it evaporates.

Artificial Poultry Hens for Rearing Chickens.

Where poultry breeding is carried on as a commercial undertaking, and where it is intended to rear the greatest number of chickens with the least number of hens, and this without interfering with their laying, artificial mothers are of the utmost importance.

The functions of a hen towards her chickens consist of forming a covering to prevent the natural heat of their unfledged bodies from cooling; also to break into small pieces any food that is too large for them; and lastly, to protect them against danger. Now, my artificial hens not only do all this, but they perform the duties a great deal better, and with less casualties to the chickens.

Most writers on poultry do not believe in artificial hatching or rearing; yet might they as well doubt growing tropical fruits and plants in England.

Chickens do neither require artificial heat nor that of their mother; all that is necessary is to provide them with a suitable covering of their bodies until they are full fledged, to preserve their natural heat, the same as with infants. During cold weather, however, their homes

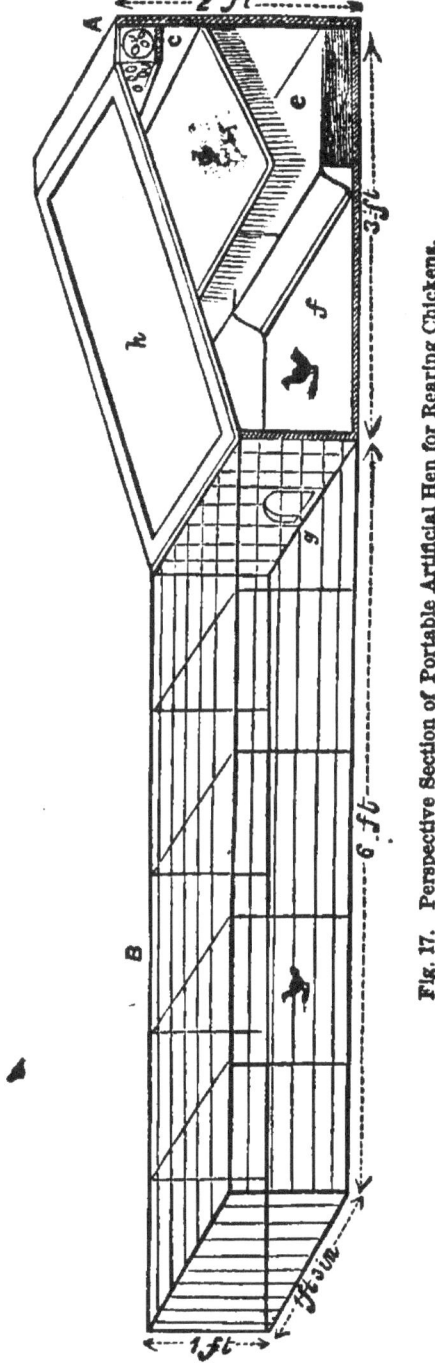

Fig. 17. Perspective Section of Portable Artificial Hen for Rearing Chickens.

must be warmed the same as for full-grown poultry; then a good ventilation without draught, a dry floor, sun light, and a small run.

The portable artificial mother, particularly recommended to breeders and amateurs, is shown by fig. 17. She performs her duties towards her chickens with far greater success than a hen possibly could do.

Reference to Perspective Section of Artificial Hen.

A is a glass-covered frame three feet long, fifteen inches wide, two feet high at the apex, and twelve inches at the rise of the glass frame. This forms a dry run in wet and cold weather. *c* is an air-flue across the frame for the necessary ventilation, and formed of perforated zinc. At each end of this flue a ventilator is fixed, by which the admission of air can be regulated according to the temperature of the atmosphere. It will be apparent that chickens are not exposed to draught by this arrangement of ventilation. *d* is a frame lined with long fleece, under which the chickens will roost the same as under the wings of a hen, and will even prefer the artificial mother, as I have ascertained by experience. *e* is about one inch deep of ashes, which may be sprinkled over with flour of sulphur: they make a dry and warm footing, and retain the heat; but they should be renewed or sifted once a week. *f*, the floor, should be slightly covered with sand and renewed every day. *g* is a small door, communicating with the open run. *h* is a glass frame, made to open by means of a slide or by hinges.

B is the movable open run, six feet long, fifteen inches wide, and twelve inches high. It is made of galvanized iron wire, which not only keeps the chickens from danger, but also prevents them from roaming. The artificial mother being portable should be taken in-doors every afternoon during the cold weather, and in the daytime should be placed on grass or dry land. However, for large breeding establishments, the arrangements would be different, and are explained in the "Artificial Rearing Home."

Artificial Rearing Home.

In poultry breeding establishments of any magnitude the portable artificial mother could not well be used with advantage; its cost, and the labor that would be required for a proper attendance on the chickens, are obstacles which cannot be overlooked without loss to the breeder; in fact, as I have stated before, in any large establishment a judicious arrangement for saving labor and for performing the work systematically by subdivision of labor, becomes of the utmost importance in a commercial point of view. Although the principles of the portable mother are strictly retained in the arrangement of the rearing home, yet it will be seen that where many thousand chickens have to be attended to in separate compartments containing not more than twelve each, the building, as shown by fig. 18, must necessarily facilitate the work of cleaning, feeding, warming, and general supervision.

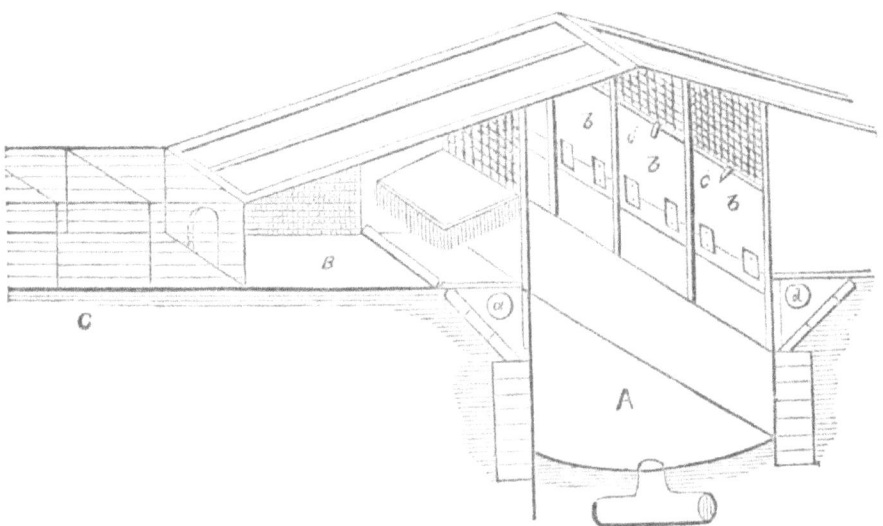

Fig. 18. Artificial Rearing Home—Perspective Section.

REFERENCE TO PERSPECTIVE SECTION OF ARTIFICIAL REARING HOME.

A is a sunk passage lined by brick walls, the floor formed of concrete, with a provision for drainage; along the whole length of this passage hot-water pipes should be fixed immediately under the roosting-place *a*. A door communicates with the covered run *b*, and wire netting is fixed over the door *c* for ventilation. The roof of this passage can either be glazed or formed of boards covered with asphalted felt, but provision must be made for an efficient ventilation. This passage should be about five feet wide between walls, and six feet high.

B is the glass-covered run; it differs from the portable hen only in this, that here the sides are formed of galvanized iron wire, and only the front is made of boards; the floor is made of concrete, covered with gritty dust. This run can also with advantage be made a little larger, say four feet long, eighteen inches wide, and two feet six inches high.

C is the open run; the floor can be formed of concrete or gravel, with an incline towards a gutter for quick drainage. The sides and top can be made of galvanized iron wire, and on the same plan as shown in the poultry home (fig. 6).

Artificial Vermin Nursery.

This is a most useful department in a poultry breeding establishment, as it will supply the poor prisoners with those dainty little morsels which in their free state they will never tire to look after.

It is well known that from the chicken to the old hen they prefer insects and worms to any grain — in fact, fowls are omnivorous, but their carnivorous appetite predominates, and they would very soon become unfit for human food were they indulged in their predilection; only in a free state they have to perform hard work in their search of insects and worms, of which, after all, they find but a scanty supply; it would, therefore, not be advisable to give fowls in a confined state too much of animal food, but only in such quantities as will prove a stimulant without injury to their health.

The effect on a fowl fed too freely on animal food becomes soon perceptible; she will pull out her feathers, and even peck her flesh until the whole of her upper body is one mass of raw flesh.

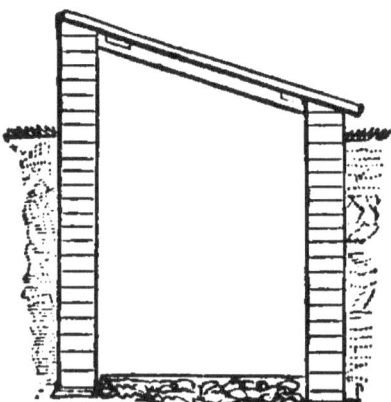

Fig. 19. Section of Vermin Pit.

It is not intended that vermin should replace the mince meat in the food for poultry, but it should occasionally be given in addition as dainty morsels in wet or cold weather.

The vermin nursery is formed of a succession of pits with concrete bottoms and brick-lined sides; the top is covered with a trap, to prevent the rain entering, which might kill the vermin. (See fig. 19.)

To propagate vermin, put in alternate layers of mould and vegetable and animal matter, such as horse dung, garden refuse, entrails of animals, dead animals, blood, &c., until the pit is filled. In a short time fermentation will commence, and the mixture will soon be converted into a living mass of vermin. If the fermentation take too long, it may be hastened by watering. In winter it is well to cover the mixture with horse manure, which will keep the vermin warm and alive.

This process of obtaining vermin is inexpensive, and it will be found very serviceable in winter for young chickens, and for stimulating the fowls to lay.

Improved Fattening Pens for Cramming Poultry.

These fattening pens are so constructed that they can be placed in the open air, forming a building of themselves.

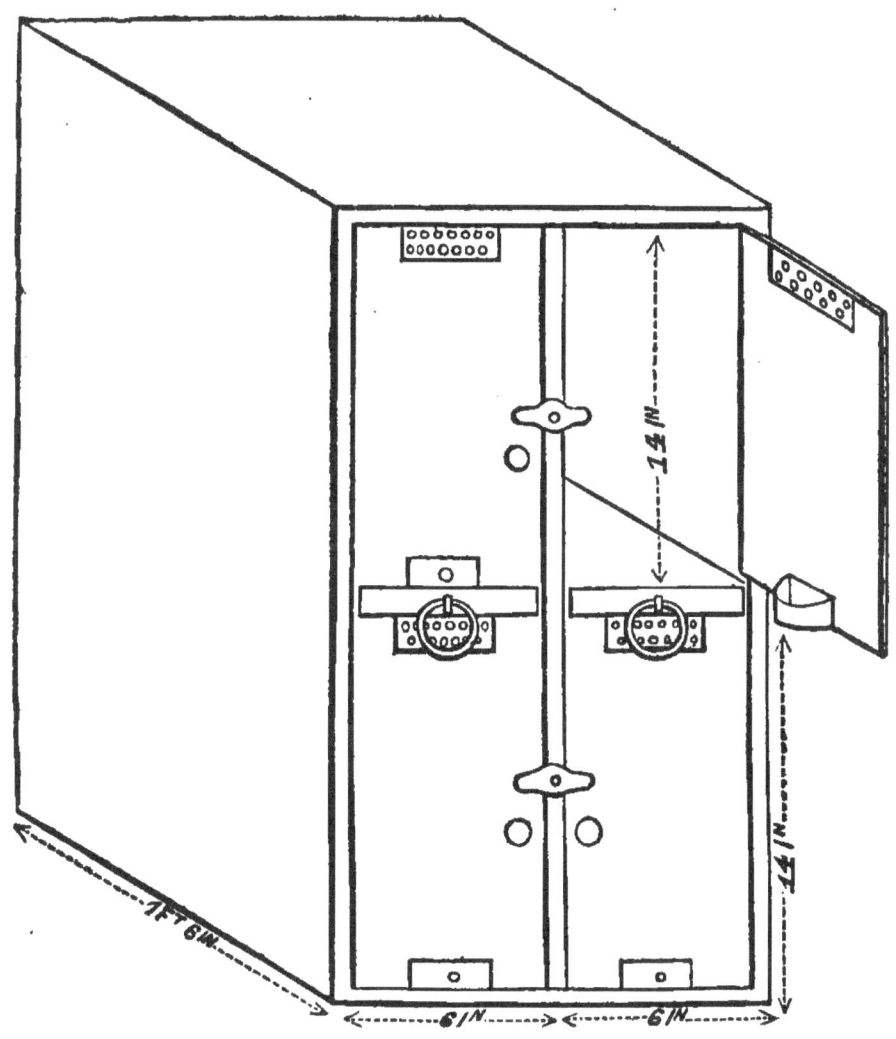

Fig. 20. Perspective Elevation of Improved Fattening Pens for Cramming Poultry.

Each fowl has her own compartment, and is thus placed in solitary confinement, and without being able to see other fowls, which accelerates considerably the fattening.

The floors of the cells should be drawn out daily, and cleaned and whitewashed; they must then be returned with the dry underside uppermost, and some sand sprinkled on. The cells should also be whitewashed for every fresh occupant.

The doors are solid boards, with a piece of perforated zinc for ventilation at the top, and a drinking-cup at the bottom. These pens combine all the sanitary requirements for the speedy fattening of fowls.

Preservation of Eggs.

Much has been written about the preservation of eggs, and many are the suggestions, but none have as yet given satisfaction, and for the sole reason that the structure of the egg is not considered in relation to the physical and chemical laws which govern evaporation, permeation, and putrefaction. The shell of the egg being porous, to admit air to the chicken during the process of incubation, allows also part of the liquid to evaporate, and the air to permeate when they are not used soon after being laid, and the air acting on the animal matter produces early decomposition and putrefaction, particularly so in a fecundated egg, in which the germ is first decomposed. Clear eggs, the produce of hens who have not been with a cock, keep fresh much longer. This can easily be exemplified by putting an old fecundated egg and an old clear egg under a hen whilst sitting, when it will be found that after the twenty-first day the fecundated egg is putrid, and the clear egg fit for use. To exclude the air from the egg, and to prevent the evaporation of its liquid, it has been proposed by some writers to pack the eggs in salt, lime, bran, saw-dust, &c., by others to keep the eggs immersed in lime-water, in salt water, or both combined; others, again, suggest to varnish or oil the eggs, and some even to parboil them.

There can be no doubt that, were the object to be accomplished solely to preserve the eggs from getting putrid, some of these suggestions might be employed to advantage; but there is more required than simply to preserve the egg from putrefaction; for instance, for kitchen use, and the breakfast table, eggs ought not only to be preserved fresh, but also free from any foreign flavor, such as lime, salt, bran, saw-dust, varnish, and oil must necessarily impart to the egg through its porous shell; and as for breeding from such preserved eggs, it is out of the question. Who has ever seen any chickens hatched from salted or mouldy eggs, or from such as have been varnished or oiled, which latter process stops up the pores through which the air, so indispensable to the formation and development of the chicken, must be admitted?

Now, the most effective, simple, and economical plan for truly preserving eggs, and without imparting to them any foreign flavor, or rendering them unfit for hatching purposes, is to use the patent stoppered glass jars, with vulcanized India-rubber joints (see fig. 21), and proceed thus : —

Immediately after collecting the eggs, put the jar in hot

water, and when thoroughly warm, so as to rarefy the air, place the eggs in the jar, the pointed end uppermost, and pack and line with paper shavings or cocoa fibres to prevent them from breaking; then close the jar before taking it out of the water, and it will be found that eggs preserved by this method will be fit for hatching twelve months after, and that those intended for the breakfast table will be as fresh as on the day when laid.

Fig. 21.

WHITEWASH.

A large quantity of whitewash will necessarily be required for sanitary purposes, but if prepared as follows, it will possess the advantages of preventing the wood from taking fire or from decaying.

Dissolve in warm water sulphate of alumina (alum), sulphate of copper (blue vitriol), and mix with the whitewash.

LIME WATER

Is most beneficial for an occasional drink to fowls; it is a preventive of many diseases, and assists the formation of bone and eggs. Prepare as follows:—

Pour over quicklime some warm water, and when the lime is slaked and settled, draw the clear water off, which can be kept for a considerable time. The lime will be useful for whitewash.

Oxide and Sulphate of Iron.

Both these can be purchased cheap from any drysalter, but they are so easily prepared that they may as well be manufactured on the establishment.

The oxide of iron (or rust) is most useful for making and improving the blood; and the sulphate of iron, a weak solution containing a large quantity of oxygen, will keep fowls lively and assist digestion. Prepare as follows:—

Take a quantity of old nails or small pieces of iron, put them in an earthen-ware vessel, then pour over them sulphuric acid diluted with water. The liquid will take up a certain quantity of iron, and form sulphate of iron or green vitriol. The rust (the oxide of iron) is obtained by mixing some diluted soda (carbonate of commerce) with the sulphate of iron. The oxide will then be precipitated, and the liquid forms sulphate of soda, which is a good liquid manure, which mix with the food or drink, as given under the heading of Food, pages 31-33.

General Plan of Buildings.

A breeding establishment on the above scale will require about four acres of land for the buildings. Six buildings, each three hundred feet long, will contain

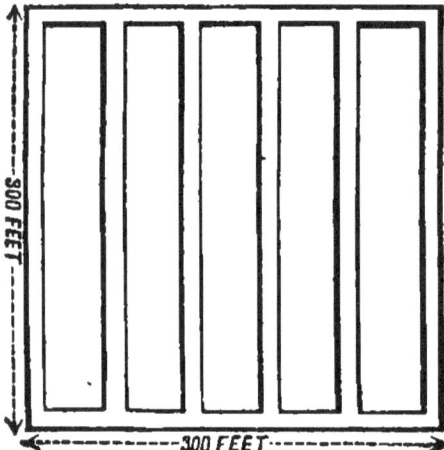

Fig. 22. General Plan of Buildings.

twelve hundred homes for poultry. (See fig. 22.) Then a building on each end, joining the six buildings, and which will be used for artificial hatching, stores, and all necessary offices. The cost of the whole will be about three thousand pounds. An uninterrupted covered communication is thus had with every part of the establishment, and the whole forms a quadrangle.

Bird's-Eye View and Section of a Poultry-Breeding Establishment.

In giving a description of the above plan for a poultry-breeding establishment, I feel compelled to notice certain observations on my system which appeared in a sporting paper. From beginning to end I clearly stated that I do not consider it possible to breed poultry profitably in large numbers on the *present system*, whatever care might be taken in a sanitary point of view; also that artificial hatching is quite of a secondary consideration, only to be resorted to during the time when hens are not broody, as I fully explained under the head of Natural Hatching, page 28.

From the above sketch it will be seen that a glass-covered passage, six feet wide, which can serve as a vinery, communicates with the poultry homes on each side. These homes consist of two runs, each twelve feet long by three feet wide; one is a closed run with gravel floor, the other an open run with horse manure. Above these two runs are two similar runs for chickens. These runs are enclosed with wire-work next to the passage and next to the field; the partition between the runs is close-boarded, so as to keep the inner run warm during winter. Efficient ventilation is provided along the whole length of the glass-covered passage immediately above the runs.

The glass-covered passage will form an excellent vinery, and this without any extra expense for building or warming; and the vines will necessarily absorb a large quantity of carbonic gas, and assist in keeping the air pure, and the soil will generate a genial, moist temperature, so essential to animal life.

The above system of keeping poultry has, moreover, many other advantages, such as —

Slow-feeding and weak fowls will be able to get sufficient food, which they cannot when a great number are fed together. The food can be supplied in the required quantity and quality to each breed or class, as it must be evident that the breeding and laying stock require a different diet to chickens or poultry intended for the market.

Each cock having only a certain number of hens allotted, they will be served better.

The means of collecting and profitably using the poultry manure.

The constant renewal of the ground will prevent it getting tainted from the fowls' droppings.

The temperature should be kept equal, and cold and dampness prevented.

Preventing diseases from exposure to cold, and wet, and contagion.

Economy in food, as poultry will eat much less when warmly housed, and deprived of roaming about.

Keeping breeds and sexes separate.

Enabling precise statistics to be obtained as to the comparative productiveness of the various breeds, and also in ascertaining what hens have ceased laying.

The early detection of hens wanting to sit.

Obtaining a larger number of eggs, and in seasons when most scarce.

A genial temperature will induce the hens to sit, notwithstanding cold weather.

This system, however, like all new systems, must be extended gradually, as old birds which have been accustomed to roam will fret and lose in appearance the first few months, but the young that are reared and fed on this system will thrive much better, and at less expense for food, than under the present mode.

The annexed sketches, one for an improved self-supplying drinking fountain, and the other for supplying poultaceous food without possibility of waste, are particularly recommended as most efficient: —

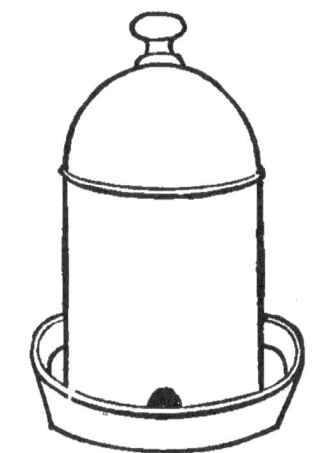

Fig. 23. Improved Drinking Fountain.

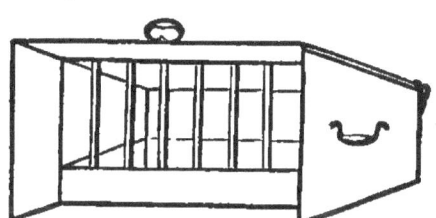

Fig. 24. Improved Feeding Trough for Poultaceous Food.

The following will be found a useful form for keeping correct statistics of a poultry-breeding establishment:—

Laying Stock.

Home No. 50 (cost price, each, 3s. 6d.), 12 Spanish, 42 weeks old.

May, 1865.	Number of Eggs.	Weight.	Hens Hatching.	Died.	Ill.	Sold.	Food.	Cost of Food per Pint.	General Remarks.
		lb. oz.					Pints.	d.	
1	10	1 8	—	—	—	—	12	½	
2									
3									
4									
5									
6									
7									
8									
9									
10									
11									
12									
13									
14									
15									
16									
17									
18									
19									
20									
21									
22									
23									
24									
25									
26									
27									
28									
29									
30									
31									

Thirty Chickens.

Home No 60 (cost price, each, 3d.), Bramahs, 10 weeks old.

May.	Food.	Cost per Pint.	Died.	Ill.	Sold.	Cocks	Hens.	General Remarks.
	Pints.	d.						
1	15	¼	—	—	—	16	14	
2								
3								
4								
5								
6								
7								
8								
9								
10								
11								
12								
13								
14								
15								
16								
17								
18								
19								
20								
21								
22								
23								
24								
25								
26								
27								
28								
29								
30								
31								

The Patent Vermin Attraction Trap.

Poultry and chickens in farm-yards are exposed to great dangers from the number of vermin which are ever ready to pounce upon them in their unprotected condition; the rat, weasel, marten, wildcat, and fox are equally destructive; therefore a trap to secure these pests, irrespective of size, has long been felt a desideratum.

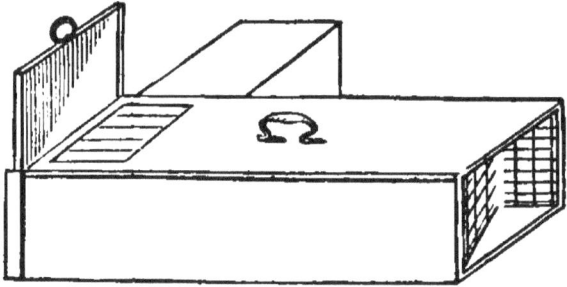

Fig. 25. Patent Vermin Attraction Trap.

From the above sketch it will be observed that the trap consists of an oblong box, the end of which draws out, and is provided with a looking-glass in the internal side, which attracts the vermin on looking in.

The entrance of the trap is formed of two spring doors made of wire, which allow the vermin to enter with the least pressure. These doors have sharp points where they meet, which, although not felt by the vermin on entering, will prevent it from withdrawing after having once introduced its head. Near to the looking-glass a bait is suspended, and a cage is also fixed with a chicken to serve as a decoy. These traps are self-setting, simple, inexpensive, fit for all sizes of vermin, and safe for the house, farm-yard, or game preserve.

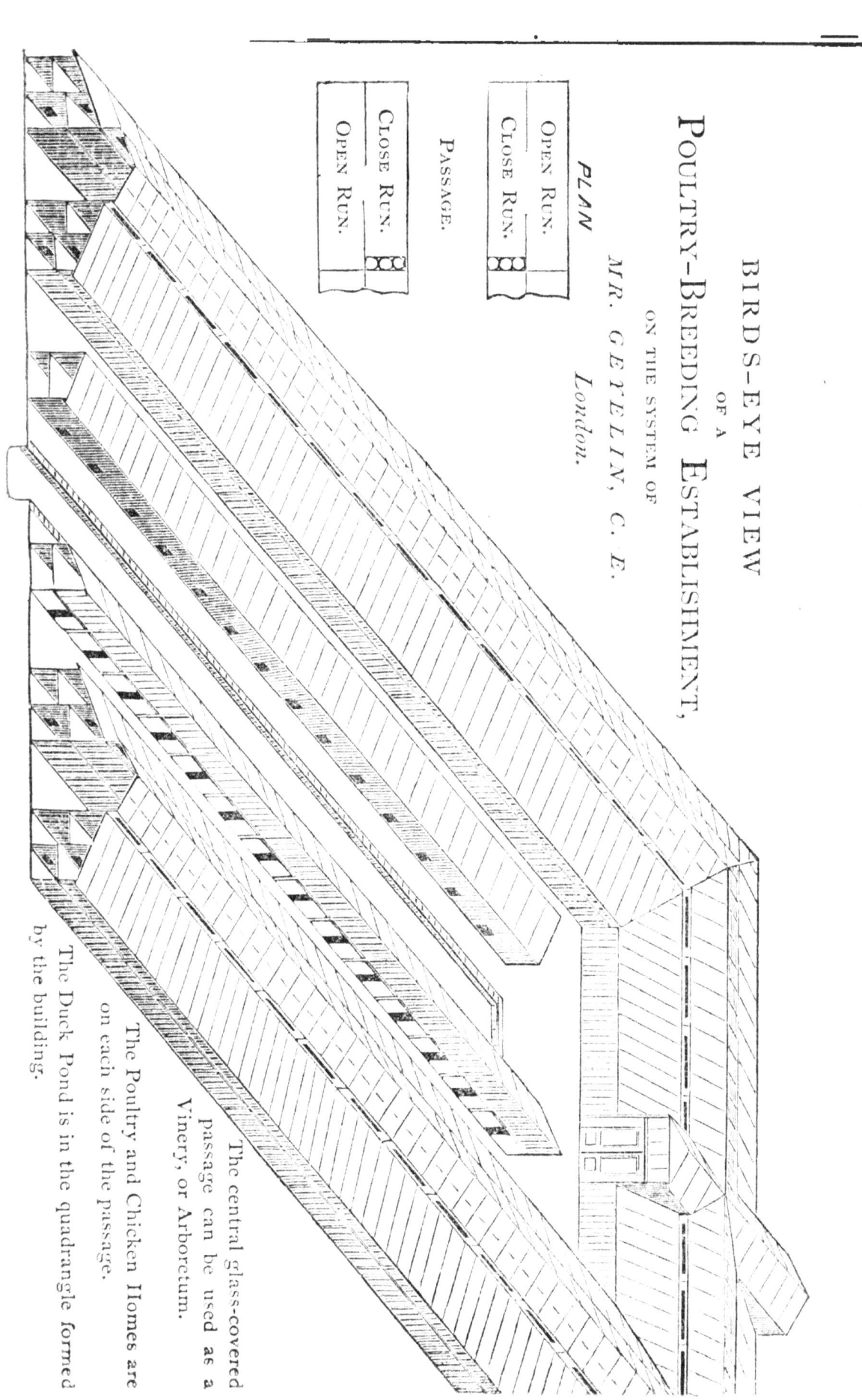

ESTIMATE OF REVENUE AND EXPENSES
FOR
A POULTRY-BREEDING ESTABLISHMENT OF 3000 STOCK FOWLS,

At an Outlay of about £3000 for Building, Fittings, and Stock.

BREEDING STOCK OF 1000 HENS AND 160 COCKS.

(*Six Hens and One Cock to a Home.*)

REVENUE.	£	s.	d.
150 eggs from each hen, or 150,000 per annum,[1] at an average price of 3d. each,	1,875	0	0
1000 broods of 12 chickens each, or 12,000 chickens at 3d. each,[2]	150	0	0
Manure at 1s. each per annum,	58	0	0
	£2,083	0	0

EXPENSES.	£	s.	d.
Interest of capital, rent, fuel, and sundries,[1] at 3s. 6d. each,	175	0	0
Food, including share of cock, 4s. 6d. per annum each,	225	0	0
Labor,[2] 1s. 6d. each per annum,	75	0	0
	475	0	0
Net profit,	1,608	0	0
	£2,083	0	0

[1] Of the above 150,000 eggs, 60,000 will be reserved for the establishment, and 90,000 sold for hatching.

[2] Is a charge for the food and time of a hen whilst sitting.

[1] Is necessarily higher than in the other classes, as the expense of a home is divided over seven birds only.

[2] Labor is also higher for the same reason.

LAYING STOCK OF 2000 HENS.

(About Twelve Hens to a Home.)

REVENUE.	£	s.	d.	EXPENSES.	£	s.	d.
180 eggs from each hen, or 360,000 per annum, at 10s. per hundred, when preserved,	1,800	0	0	Interest, fuel, rent, &c., 1s. 6d. each,	150	0	0
2000 broods of 12 chickens each, or 21,000 chickens at 3d. each,	300	0	0	Food, 4s. each,	400	0	0
Manure at 1s. each per annum,	100	0	0	Labor, 1s. each,	100	0	0
					650	0	0
				Net profit,	1,550	0	0
	£2,200	0	0		£2,200	0	0

CHICKEN CLASS, 50,000 PER ANNUM.

(Six Months old, and fattened for the Market.—About Twenty-four to a Home.)

	£	s.	d.		£	s.	d.
50,000 chickens, at 2s. 6d.	6,250	0	0	60,000 eggs at 3d. each, and 3d. each for hatching ditto as above, 6d.,	1,500	0	0
Feathers from ditto, 1d. each,	208	0	0	Interest, fuel, rent, and sundries, 3d. each,	624	0	0
Manure from ditto,	624	0	0	Food, at 9d. each, for six months,	1,872	0	0
				Labor at 2d. each,	416	0	0
					4,412	0	0
				Net profit,	2,670	0	0
	£7,082	0	0		£7,082	0	0

10,000 eggs are allowed as clear eggs, which will, however, be valuable as the food for chickens.

ARTIFICIAL HATCHING.

REVENUE.	£	s.	d.	EXPENSES.	£	s.	d.
30,000 eggs, at 3d. each, as charged in chicken class,	£375	0	0	At 1d. per egg,	125	0	0
				Net profit,	250	0	0
					£375	0	0

The entire profit from the sale of choice birds from among 50,000 per annum may safely be computed at £1000.

The extra profit from contracts for the supply of the navy, clubs, hotels, and families, may also safely be set down at £1000.

From the above it will be seen that ample allowance has been made for expenses, and that the revenue will bear disappointment and still leave a very large profit in proportion to the capital employed.

EPITOME OF REVENUE AND EXPENSES.

	£	s.	d.		£	s.	d.
From eggs,	3,675	0	0	For interest, &c.,	949	0	0
" Natural hatching,	450	0	0	" Food,	2,497	0	0
" Artificial ditto,	375	0	0	" Labor.[1]	591	0	0
" Chickens,	6,250	0	0	" Artificial hatching,	125	0	0
" Manure,	782	0	0	" 60,000 eggs, and hatching,	1,500	0	0
" Feathers,	208	0	0		5,662	0	0
" Choice birds,	1,000	0	0	Or a net profit of,	8,078	0	0
" Contracts,	1,000	0	0				
	£13,740	0	0		£13,740	0	0

[1] The sum charged for labor will admit of an efficient staff of at least thirty hands, young and old, of both sexes.

THE LAWS OF NATURE

In Relation to Poultry-keeping from a Commercial Point of View; and Answers to Questions.

In my Preface I stated that I should confine myself exclusively to giving publicity to such facts as I have proved by actual experience. My knowledge of the world cautioned me not to introduce anything which might savor of theory, particularly to a class of readers who undoubtedly by their education are conversant with the laws of nature, on which facts are based. I was, therefore, but little prepared to have so many questions to answer, which more or less compel me to do that which I endeavored to avoid in fear of being considered pedantic. Though I feel flattered by the great interest my treatise has created, and though an explanation of the laws of nature will prove interesting to many, yet I trust that my correspondents will not consider it a want of deference on my part if I abridge as much as possible my explanations, but still with a due regard to giving a satisfactory reply to all inquirers.

Egg Preserving.

1. *Question: Does it make any difference to preserve eggs a few days old?*

The egg comes from the hen at blood-heat, the liquid then fills every part of the shell, gradually the egg cools, and the air contained in the egg is condensed, thus leaving a vacuum. Now, as the shell is porous, and the pressure of the outer air much greater, it forces itself gradually through the pores of the shell until the equilibrium is reëstablished, thus forming the depression of the fluid part observable in old eggs at the round end; and as the contact of the air with the fluid part very soon alters the taste, and renders them unfit for hatching from, it becomes essential that the eggs should be preserved as early as convenient after being laid.

2. *Why should eggs be preserved better in rarefied air than by merely packing them in air-tight jars?*

The variations in the temperature of the atmosphere from below freezing point to summer heat are important considerations in preserving eggs. The elasticity and expansive properties of air need not be explained here, as they will require a full explanation under the questions in reference to regulating heat. I will, therefore, only say, that if an air-tight jar were closed up during cold weather without the air within being first rarefied, it would, provided it remained air-tight, stand a good chance of bursting during the summer heat, which would expand the air in the jar, and the pressure on the eggs would be so great that a quantity of air would be forced on the fluid through the pores of the shell. Were it possible to preserve the eggs immediately on being laid at the temperature of blood-heat, and during the hottest summer days, the jars would not require rarefying; but as such conditions are almost impossible to command, as the eggs must unavoidably on cooling absorb a certain amount of air, and as the atmosphere might become still warmer than on the day of filling the jars, it becomes necessary to rarefy the air in the jars even in summer, although not to such a degree as during cold weather. The air in the jar being thus rarefied, its permeation through the shell will not only be prevented, but the excess of air in the egg will actually be withdrawn until the equilibrium is reëstablished.

3. *Are the eggs not affected by the heat whilst being packed?*

Although the jars are placed in boiling water, the internal temperature never exceeds blood-heat, and as eggs are not affected by that temperature, which is the same as that at which they come from the hen, it becomes only necessary to avoid packing the eggs against the jars without a lining of cocoa-nut refuse, which ought to be perfectly dry, and used warm for packing.

4. *How can you tell when a jar is air-tight?*

To ascertain whether a jar is air-tight when empty is easy enough; it is merely necessary to fill the jar with boiling water, and when thoroughly warm to empty out the water, then close the jar quickly with the air-tight cover, and place it inverted in a tub of cold water. The air in the jar thus rarefied will be condensed by the cold water. If not perfectly air-tight, some water will find its way into the jar, which can be ascertained by opening the latter a few hours afterwards. This test, however simple, cannot be applied to filled jars, as it would be necessary to open the jars again. Now, this very same question I put to several pickling firms, and obtained the unsatisfactory reply that they consider when jars are air-tight when empty they will be equally so when filled. This, however, cannot be depended on, as the cover may fit in one place and not in another, or it may not be screwed down so much at one time as another.

5. *Can you at any period ascertain whether the jars continue air-tight?*

6. *Which are the best air-tight jars, glass or stone ware?*

These two questions I will endeavor to answer under one head; and to prevent my being considered to advocate one principle more than another from an interested motive, I must inform my correspondents that to most scientific men and poultry-breeders it is well known that egg preserving has formed the study of some of the most eminent chemists in Europe, and that until I published, through *The Journal of Horticulture*, my simple and yet the only truly effective mode of preserving eggs for any length of time, no satisfactory means had been discovered. The intense interest this discovery has created throughout England has induced me to ascertain which of the professed air-tight jars are really so, in order that the public may not lose their confidence in so important a discovery on account of the jars not being to be depended on. Through the kindness of an eminent firm in the pickling trade, I have been enabled to make experiments with the various so-called air-tight jars, few of which really were so, and all without exception objectionable on account of their construction. Then there came another important consideration: how can it be ascertained, when the jars are filled, whether they are air-tight, and how long they will remain so? This, of course, was a perfect impossibility with the jars as at present manufactured.

These important deficiencies in air-tight jars for preserving eggs have led me to invent a jar purposely for egg preserving, and which jar is not only perfectly airtight, but it will show at a glance whether it is so, and how long it remains so, by means of its patent pneumatic self-indicating cap. I have every reason to believe that these jars will remain air-tight for any number of years, and that the eggs preserved in them will remain as fresh as on the day they were laid, and fit for hatching and the breakfast table. Now, although such jars can also be made of glass, which would have the advantage of showing the eggs, yet when it is considered that glass is liable to crack when put into boiling water, particularly during cold weather, it is my opinion that glazed stone ware is preferable.

PATENT PNEUMATIC SELF-INDICATING AIR-TIGHT JARS.

Fig. 26.

From the above sketch it will be seen that the jar has two covers; the inner is made of stone ware, with a ridge parallel to the neck of the jar, between which plaster of Paris or cement is poured, which soon gets hard and secures the lid, which prevents the eggs being jolted during transit; the hole in this cover is to maintain the equilibrium of the air between the two covers and the interior of the jar. The outer cover is formed of the best India-rubber, with a strong ring of the same material, which fits in a groove. The mouth of the jar is four inches in diameter, which allows of quick packing.

Packing the Eggs.

Place the jars in boiling water for about ten minutes; then plait a layer of well-dried and warm cocoa-nut refuse on the bottom, and pack the eggs, taking care not to let them come in contact with the sides of the jar; as soon as the jar is filled, put on the inner cover, and pour some cement or plaster of Paris between the ridge and sides of the jar, then draw an India-rubber cap over the mouth of the jar, after which it should be immediately withdrawn from the hot water and immersed in cold water, which will condense the rarefied air in the jar, when it will be found that the difference of the atmospheric pressure is at least ten pounds to the square inch, which forces the India-rubber down to the inner cover; and as long as it remains so it will be a sure index that the jar remains perfectly air-tight. The inner temperature of the jar, although placed in boiling water, will be found not to exceed blood-heat, which being the same at which the eggs come from the hen, they cannot be affected.

Why Eggs should be packed with the Small End upwards.

This advice is so opposed to all published directions on the subject, that I feel bound to give my reason for it. Most persons will have observed that when an egg is boiled a vacuum is observed at the round end, which is more or less extensive according to the age of the egg; now, this is on that part of the egg where the shell is most porous, and where the air is admitted most freely. This air chamber is of the utmost importance to the chicken whilst hatching, as it serves to equalize the supply of the necessary air under the variations of the outer temperature; and it will be found that the chamber gets enlarged as the hatching proceeds; therefore, when eggs are packed with the small end upwards, the liquid presses on the most porous part of the shell; consequently, for the air to penetrate the egg, it would have to lift the weight of the fluid.

Warming Poultry Homes.

On this subject I have had many inquiries. I will therefore endeavor to give my reasons why I prefer hot-water pipes fixed immediately under the roosting-perches. Where stoves or open fireplaces are used, the heat is concentrated; therefore the cold air rushes from all parts to that particular spot, which cold draughts are most injurious to the health of not only poultry but all animal life; in fact, these cold draughts are the cause of most diseases in England, where persons sit before an open fireplace, and right in the current of the cold air. Consumption, and colds of all description, could be considerably averted by a different mode of warming our houses; but, though the hot-water system is not applicable to our private dwellings, it is eminently so for a poultry-house. The heat from the pipes is equally radiated from all parts of the building, and the pipes being immediately under the roosting-perches, it will not require the maintenance of so high a temperature as when the pipes are near the floor, where not so much warmth is required, as it is well known that whilst the body is in motion the natural heat is sufficient to keep us warm, but that as soon as the body is at rest or asleep, the circulation of the blood becomes less active; consequently it cools sooner, as is evidenced by the fact that the clothes that keep us warm in action are not sufficient to protect us from cold during sleep. There is another consideration why I prefer the pipes under the roosting-perches: it is because, being placed at mid-height of the roosting-room, they are between the heavy and light gases which are necessarily generated in all places where animal life is congregated.

Our System of selling Poultry.

This subject is of such vast national importance that it deserves the most serious consideration of all who take an interest in our national welfare; it is not only on account of the immense sums we annually pay to foreigners, but also on account of the enormous destruction of poultry, which, under a different system of sale, would become available for the people at a price to compete with butchers' meat.

There is no country under the sun where obsolete customs or protective prices have been so successfully replaced as in England, and this against the most ominous prognostications. Need I enumerate the long list of articles, from the postage-stamp to tea?

Up to this very day we are informed that the supply is always equal to the demand, even in poultry. Has it not always been so since the beginning of the world? But this is not the question. What is the result of our free-trade principles, our improved machinery, our improved agriculture? Why, a hundred-fold consumption of these very articles which were then, as now, said to be supplied according to demand; a participation in the comforts of this life by the poorer classes; a steadier and more remunerative employment both of capital and labor. True, but there are no protective duties on poultry; it is simply a question of price between dealer and customer. By appearance this looks fair enough, and the manner of sale is so old and deeply-rooted that it is accepted both by breeder and the public as a perfectly fair way of dealing; yet who would ever purchase a leg of mutton, or a surloin of beef, at so much apiece? Why, the very poulterer who feels insulted if any person asks him what the fowl weighs would no more think of buying a joint by guess than he would of selling a fowl by weight.

Next you will be told that poultry must always remain a choice morsel for the upper classes only, as the poor will never be able to afford the price; that the production is, after all, limited, and that the climate of England is not suitable to cheap poultry breeding. Now, these assertions are based on mere narrow-minded prejudices; there is no climate in the world more favorable to animal life than that of England, as is proved by our statistics; the very dampness of our climate prevents those extremes of cold and heat from which more southerly countries suffer; and nowhere can fowls be produced to compete with ours in size and flesh; and all travellers will agree with me that large-sized fowls can only be found in damp regions, while those of dry and hot soils, such as Egypt, are comparatively small. As regards a limited production, I have shown in my treatise on "Poultry-keeping from a Commercial Point of View" that fifty thousand fowls can be reared per annum on four acres of land, and at highly remunerative prices, and much below that of butchers' meat. Poulterers will, moreover, maintain that poultry is exceptional; that it cannot possibly be sold by weight; that the price must necessarily depend on age, breed, quality, and feeding, and cannot be classed. Believe it not. Have we not beef, mutton, pork, &c., from fourpence to a shilling per pound, according to quality? Is not every produce now sold, according to its intrinsic value, by weight or measure? And why should poultry form the exception? I am very much mistaken if tne public will not be able to purchase chickens at the price of Ostend rabbits within two years. Let the public once see tickets in shop windows, prime chickens at sixpence per pound, or at any other price, according to quality, and you will find the commendable wish of Henry IV. of France realized, that every family shall be able to have a fowl for their Sunday dinner. Why, it will create such a revolution in the national production and consumption of poultry as the world never witnessed before. People would then begin to understand and appreciate the value of poultry, which, up to this day, is kept either merely as a matter of fancy or a necessary adjunct to a farm-yard, but whose productiveness is disputed by many.

What is the result of our present system of selling poultry? Why, that tons of poultry are annually destroyed because sufficient customers cannot be found to pay the fancy prices. This system is neither fair to the breeder, the dealer, or the public. The first consign a quantity of poultry to a salesman, and obtain but a poor return; the second's percentage is naturally in keeping with a limited sale, and the public are obliged to pay fancy prices, or forego an article of food which ought to be within the reach of even the mechanic. As the first commercial nation in the world, we ought not to purchase food in foreign markets when we can produce it at home. At the present time poultry is collected from all parts of the country by higglers, who consign it to a salesman, from whom the poulterers purchase at so much a head: sometimes the demand is good, and fair sales are effected; other times the supply is too great; then the poultry past keeping is sold at a nominal price to costermongers (rather than have it condemned by the market inspector), who, in the garb of countrymen, hawk it about the suburbs of London in a state unfit for human food.

By the immense importation of eggs and rabbits, foreigners have shown us how to proceed to alter our system of selling poultry. When they found that poulterers would not agree to their terms, they made arrangements with cheesemongers, dairies, chandler-shops, and others, and the result is, that in season the people can now purchase twenty-four eggs for a shilling, rabbits at sixpence per pound.

If, therefore, it pays the foreigners to collect these articles of food abroad, to pay carriage, freight, agency dues, and all other expenses connected with such vast importations, does it not seem passing strange that we, as a nation, do not even try to see what we can do for ourselves?

Now, what I propose is this: that when once a poultry-breeding company is formed, that the directors should invite coöperation from poultry-breeders in general, and establish an agency in all the principal towns for supplying poultry to such shopkeepers only who will undertake to sell it by weight; this will now be of no great difficulty, as those who sell rabbits would at once undertake the sale of poultry.

EXTRACTS

FROM THE "JOURNAL OF HORTICULTURE AND COTTAGE GARDENER."

Home Supply of Poultry and Eggs.

How long does it take for a question to go through all its phases, to settle down as a recognized fact before the public, with the certainty that it will only be disturbed now and then, at stated periods, to be ventilated according to some, or to have the accumulated dust of years rubbed off according to others? Fourteen years ago, the public one day recognized the fact, that poultry was a pursuit, that it was deserving of encouragement, and then some thought it was a mania. The pursuit of the trade of a "poulter" is not one of yesterday. Its Guild or Company ranks among the ancient ones of the city of London. It has in its day lent money to the Sovereign of the United Kingdom. It still exists, and has its chartered rights; and its bequests and benefactions go back to the sixteenth century. It seems now as if its claim to a share in providing food for the vast populations in the metropolis and large towns were about to be properly considered. We are no longer self-supporting in the way of food. Many of us can recollect in one of the old Anti-Gallican songs it was said,—

> They want to get our flesh and blood,
> Our beef and beer.

Things are altered — we get a good quantity of theirs; and half the continent is laid under contribution to supply our carnivorous propensities. Good sound men of figures prove that the supply of food decreases; others point to an increasing population, and the increasing price of meat. The leading journal of the world but lately called attention to the figures of our imports in the way of food, and was obliged to pause at one item. "A million of eggs imported for every working day in the year;" turkeys by thousands; rabbits by the ton. These are helps to the food necessary to feed the metropolitan millions; but the question naturally suggests itself, Do we do all in our power to provide more of these things at home?

The egg trade with Scotland is becoming a very large one. We import eggs from France, Holland, Belgium, Switzerland, and part of Italy. Cannot we do something towards providing ourselves with these valuable luxuries, and thereby not only increase the number, but probably decrease the price of them? We have in our favor, and, consequently, to our profit, all the expense of foreign agents, of travelling, freight, carriage, and dues. With eggs produced at home, nearly all this would be avoided. If those who can keep poultry will not keep them, then those who have the inclination without the convenience must endeavor by association to find out the means of carrying out their theories and ideas on the subject.

We believe we are correct in stating that plans are now afoot which will, in all probability, result in calling public attention to the subject, and in giving the question a fair trial on a large scale.

Poultry and Egg-preserving Company.

Whatever differences of opinion may exist as to whether poultry can be kept profitably in England from a strictly commercial point of view or not, it is certain that this subject will never be satisfactorily decided by any amount of mere theoretical assertions *pro* or *con;* nor will the problem ever be solved in a national point of view by the success of one or more private persons, whose balance-sheet would be discredited by many sceptics, as we have had ample evidence, in improved farming, the results of which were published year after year by Mr. Mechi and other pioneers. Moreover, to render poultry breeding profitable in England, it must be evident to most persons that the system cannot be carried on as it is now; also that the trial must be made on a somewhat extensive scale to allow of growing or purchasing food at a cheap rate, and of a subdivision of labor, and last, but not least, for establishing a profitable outlet for the produce. Now, there are many undertakings which cannot well be tested on a small scale in order to prove what the result would be on a large one, and we believe that poultry breeding in England is one of them.

It is too much to expect that any private gentleman would run the risk of an outlay of some thousands of pounds to ascertain the value of a new plan of breeding poultry in order to benefit the nation; but as the subject is really one of national importance, it is highly desirable that it should command a fair trial. This can be best accomplished by a public company, whose accounts would be audited by independent accountants, and the statistics of which would be reliable, and, consequently, valuable to the country at large.

In a former number it was stated that plans were afoot which would, in all probability, result in calling public attention to the subject, and in giving the question a fair trial on a large scale. Since then the preliminary prospectus and the plans have been issued.

Mr. Geyelin has evidently the utmost confidence in the success of the undertaking, as he has taken on himself the trouble and expense of bringing this subject fairly before the public; and he gives the free use of his inventions to the Company, not for a cash consideration, but for shares the value of which must entirely depend on the profitable result of his system of poultry keeping.

A Company of this description requires only half a dozen gentlemen earnest in their efforts to carry out the proposed system. On our part we shall watch the result of this long-vexed question, Can poultry be bred profitably in England? with great interest, for if successful, poultry will be sold, as it ought to be, by weight, like other articles of food.

Home Supply of Eggs and Poultry.

Is there any valid reason why England should not supply her own wants in the shape of eggs, poultry, and rabbits? I dare say the money we pay foreign countries for these necessaries, does not fall far, if at all, short of five hundred thousand pounds annually. Can they not be produced as cheaply, abundantly, and profitably at home as in France and Belgium? Is it not time that some efforts should be made to solve this problem? I am aware much has been done for the last few years to improve the breeds of our poultry, but I have never heard of the production of eggs and poultry having been attempted in a large way, as a matter of trade or business, though I have often been told that to make this stock pay they should be kept in such numbers as to employ the whole time and attention of working people. M. de Lavergne estimated the value of poultry in France at eight million pounds, while that in England was no more than eight hundred thousand pounds.

As a national branch of rural economy, we know nothing in England of the breeding and management of poultry: hence practical men never think of embarking in a pursuit which is found so profitable in other countries.

We sadly want sound, reliable, practical information on this subject, and if through your columns some of your correspondents will endeavor to ventilate this question, much public good may be the result.

If one acre of average land were cropped with the grain, pulse, and roots, most suitable for feeding poultry, how many heads should it maintain for one year? Again: What might be a fair moderate profit to expect per thousand in keeping poultry thus on a large scale, assuming suitable houses, warmth, care, and ventilation for such stock?

I have heard and read much on the subject of artificial incubation, and I knew a lady who produced all her own poultry by a most ingenious incubator of her own invention; but I never could ascertain how far the system could be relied on in a commercial point of view, which is the practical test of its merit. If undoubtedly a success, then I can see no limits to the profitable production of poultry in England. Turkeys and geese of the largest breeds are now worth very nearly as much as a fat sheep of the smallest breeds, and it is passing strange that you must give two pence for a "new-laid egg," when you may buy a quarter of wheat for thirty-two shillings. High authorities tell us it does not pay to feed oxen, and farmers now say they are selling grain at prices for which it can hardly be grown; so I am induced to ask if the experiment of a regular and well-managed poultry farm would be likely to succeed; for if so I should be very well inclined to try if England cannot produce eggs and fowls as cheaply as France; and, further, if the air of our own happy land is not fully as congenial as that of Belgium to an — OSTEND RABBIT.

Poultry Keeping from a Commercial Point of View.

It is for Englishmen to determine whether England is capable of profitably supplying eggs and poultry for her own consumption; but I am painfully surprised to notice how very few seem to take an interest in the subject, which you have been pleased to bring before your numerous readers with laudable zeal and perseverance. When I had the honor of addressing you in No. 204, I was in total ignorance of the highly interesting discussion raised by Mr. Geyelin, whose ability and industry entitle him to public sympathy and support, at least to the extent of fairly trying if it be commercially possible for England to feed her own people. He goes very fully into figures to show this may be profitably done. Some few, and I regret to say but very few, of your numerous correspondents seem to notice his remarks, though none can doubt their importance. This is not very creditable to the poultry amateurs of England, who are ever ready to discuss most zealously and learnedly about the breeds of fowls, or the proper colors of cocks' legs and tails, but who seem to ignore such practical dry business details as the produce of hens and the cost of feeding them.

These are mere questions of detail, and resolve themselves into a matter of pounds, shillings, and pence; but the success of Mr. Geyelin's project depends entirely on the amount and cost of production and the market value of the produce.

It is not for me to interfere between Mr. Geyelin and "C. S. J.;" so I leave them to settle as they can, for it is evident both mean well, though differing considerably in their views. I must, however, notice two correspondents in No. 206, one "An Old Subscriber," on the wholesale price of eggs, the other "Barndoor," on poultry food and annual egg produce. Both are apparently adverse to Mr. Geyelin's figures, but in point of fact neither seems to affect him in any way. His scheme only refers to London prices, and to the productive powers of hens fed and lodged, so as to stimulate the utmost powers of production. It is but fair to Mr. Geyelin and the public to use the utmost candor and sincerity, so as to put the case fairly. He fixes the produce of hens fed on his plan at one hundred and eighty, while "Barndoor" writes,

"Cochins, Bramahs, &c., should lay one hundred and twenty eggs in a year." True, but how many more may be reasonably expected? Does "Barndoor" give this as the actual result of his own experience? If so, will he say how the hens were treated, their ages, &c.? I admit the mere opinion of one man, or even his practical experience, can carry but very little weight; I can, however, quote some well-known authorities which fully sustain Mr. Geyelin's calculations about the produce of hens, while but one goes so low as one hundred and twenty, the figure of "Barndoor," for the best known egg-producing breeds. Cobbett says eleven hens should give two thousand eggs and one hundred chickens, if well fed, in one year, and allows eighteen bushels of barley to feed them with one cock. Richardson relates that three Polish pullets laid five hundred and twenty-four eggs, cost sixteen shillings and six pence. Baxter records that four hens laid seven hundred and ten eggs one year, at a cost of one pound two shillings and a penny halfpenny, and five hundred and ninety-four the next year, at fifteen shillings and nine pence halfpenny.

In the work called "Farming for Ladies," we read, "Hens lay nearly all the year round, except when moulting and in the depth of winter; but generally speaking at least ten to twelve or fourteen dozen eggs a year may be counted on." I dare say the experience of most of your readers will differ quite as much as that of those writers, so much depends on the peculiar circumstances of each case. In my opinion a fair average can be taken only by the actual results obtained by a large number of the most careful breeders, who keep their poultry in the best and cheapest way. I am in candor bound to add, that in "Chambers' Information for the People" it is broadly laid down that no hens will pay for their food if it is all purchased.

This question is strictly a national one, for it is most important to ascertain by actual experiment if eggs and poultry may be produced by ourselves as cheaply as by various nations immeasurably behind us in everything relating to agriculture. If Mr. Geyelin can succeed in teaching Englishmen that this may be done, then few will deny his claim to be fairly considered a national benefactor. At present his task is not an easy one. As it is really the duty of every well-wisher of his country to aid this noble effort by every practical means, so your experienced readers should each contribute, as far as he can, to simplify this question by clearing up the doubts and difficulties that surround it. Men are naturally timid and shy of embarking their money in any novel experiment of which they have but very scanty knowledge; and which of us can say that he has any knowledge of thus producing eggs and poultry commercially in England? For the present I shall say nothing about the little animal which is well known to be a decided commercial success in Belgium, where many are largely engaged in breeding, feeding, and exporting to hungry, wealthy England the — OSTEND RABBIT.

POULTRY AND EGG COMPANY.

By nature I am a timid and cautious man, and dread to see my opinion appear in print; I feel, however, I must make an exception in favor of Mr. Geyelin, whose interesting articles on poultry-breeding, published in your valuable journal, I have read with much pleasure and advantage. For some years past I had an idea of breeding poultry on a large scale, but the dread of becoming the laughing-stock of my neighbors in case of failure has hitherto deterred me from doing so. When I saw the advertisement in your journal for the formation of a Poultry-breeding and Egg-preserving Company, I determined to come up to London and see Mr. Geyelin, with a view to elicit full particulars, as I consider that such a company would, if formed, supply me at a trifling risk with such practical information as I could not expect from personal experience at a much greater outlay.

Mr. Geyelin has very kindly explained to me his plans, his mode of keeping accounts, and feeding. In fact, he has given me such ample and satisfactory explanations that I feel in justice bound to say that if ever a plan appeared to me feasible it is that of Mr. Geyelin. The solution of the problem whether poultry can be bred as profitably in England as on the Continent is of national importance; it behooves, therefore, all persons who take an interest in poultry breeding to contribute to some extent towards the expense of making the experiments. I, for one, subscribe for ten shares of five pounds each; and let the result prove even a failure, I shall still consider it a good investment, as it will have been the means of obtaining at least some valuable information; but should it, on the other hand, prove a success, it will confer a great benefit on the country. There are opportunities in life which, if allowed to pass, may never present themselves again under such favorable circumstances. At the present moment there is a gentleman, not only willing, but able, to make the experiment on scientific principles, and whose statistics, if kept according to the plan I have seen, will be so precise as to become most valuable to every poultry breeder. If we lose this opportunity, shall we ever have the like again?

Shall it be said that Englishmen are so little enterprising that we prefer to purchase in foreign markets rather than ascertain at a trifling individual expense whether we cannot supply our own wants in the shape of eggs, poultry, and rabbits? — A SUSSEX FARMER.

POULTRY KEEPING FROM A COMMERCIAL POINT OF VIEW.

"Nemo," the defender of "C. S. J.," has my best thanks for his kindly lecture as to what is required to establish the success of anything nowadays. I do not find fault with his opinion as regards the profits poultry breeding will yield, and in the absence of any actual statistics of my system on a large scale, I can only reiterate what I stated in my reply to "C. S. J." There are, however, a few assertions with which I beg to differ, even with "Nemo."

1st. Were it not for sanguine minds, few improvements would ever be carried out; and were even the minimum profits given, there would still be found many persons who would, with just as good reasoning, reduce it below zero.

2d. The artificial hatching can be entirely dispensed with in my system, as for every one thousand hens, I can rear at least ten thousand chickens; and under any circumstances it will only be resorted to to hatch chickens when hens have ceased to be broody. As to whether my system of artificial hatching is superior to that of Cantelo or any other, I may state that I do not claim any originality, but rest the success only on the well-known law of a uniform temperature, at which there is not the slightest difficulty to hatch chickens; and this uniform heat can be maintained either by manual or mechanical means, which are well known to engineers, and which will be described in some subsequent number of this journal, under the laws of nature in relation to poultry keeping. The failure of Cantelo and others cannot be ascribed to the hatching, but solely to the rearing of the chickens. Now, this is my system, and on it I rest the success of poultry breeding; and though I do not intend to rely on artificial hatching, yet I shall entirely depend on artificially rearing all the chickens, whether hatched by a hen or by an apparatus.

3d. My system of breeding poultry, and its profits, can no more be judged by the present mode than railway travelling, when first projected, could from the old stage-coaches. For its success it will require a staff of servants, and a subdivision of labor, then a good disciplinarian as superintendent; and the whole will form a piece of mechanism which will work with the greatest precision, and afford such statistics as will surprise sceptics. To carry this out is a mere matter of money and will, whether by a private individual or a public company; but the idea of an association of working partners to attend to fowls is simply ridiculous, and I doubt whether, if they could be found, they would long remain a united family.

4th. If "Nemo" will take the trouble to read the article on natural hatching, No. 198, he will find that I advocate natural hatching and artificial rearing. Has it ever occurred to him, in estimating the annual profit of a hen, to charge to her credit the hatching of one or two broods? which surely is more valuable than the number of eggs she could lay in twenty-one days.

REPORT OF MR. GEYELIN,

May 17, 1865.

Gentlemen:

The nation will owe you a debt of gratitude for having by your discriminating confidence in my plan proved yourselves the pioneers to an increase of our national wealth and comfort.

The section of the intended building you have inspected to-day must have convinced you that, whilst constructed on the most economical plan, it yet combines all necessary requirements for the health and comfort of poultry, and the saving of labor.

We are not about to carry out any new invention in poultry breeding, but merely a wise combination of well-established facts; individually the facts are well known, but a combination of them applied to poultry breeding has hitherto escaped the notice of rural economists. For instance, it is well known, —

That earth is the best and cheapest deodorizer.

That poultry manure is a first-rate fertilizer.

That in moderation the gases generated by vegetables are beneficial to animal life, and *vice versa*.

That poultry require vegetables, and vegetables manure.

That poultry cannot thrive on a manure-tainted ground, which consequently requires frequent renewing.

That the earth requires manuring after each crop.

Now, when the above well-known facts are considered in connection with poultry breeding, it must become evident to the most superficial observer that to render it a commercial success, we must follow more closely the great teachings of Nature: in truth, the animal and vegetable kingdoms are so closely allied, and so dependent the one on the other, that to treat each as a distinct undertaking must necessarily increase the expense of production, and consequently decrease the profit.

I will now briefly review poultry breeding and vegetable growing as separate undertakings, in a commercial point of view, in order to show that in sound rural economy the two ought to be combined, both for sanitary and economical purposes.

Poultry Breeding.

In any establishment where large numbers of poultry are kept, the ground must often be renewed to prevent it getting tainted; this requires labor and materials. Now, there is only one material which combines all the requirements for the floor of a poultry home, and with which Providence has supplied us bountifully, namely, earth. It is composed of all the necessary materials to the animal economy of the poultry; it is of a deodorizing nature; and, when tainted with the manure, becomes a valuable fertilizer; but even earth can absorb but a moderate amount of decaying matters without losing its valuable properties; and, in this again, Nature teaches us that what is beneficial in moderation becomes injurious in excess. Then comes the question of a cheap supply of earth, and of the disposal of the tainted without causing a nuisance by its accumulation. In this, also, Nature comes to our assistance: we know that by growing vegetables in manure-tainted land they absorb and feed on the noxious gases, and give out in return oxygen gas, so essential to the health of animal life. Thousands of tons of manure, decaying vegetable and animal matters, are annually buried in the earth; and yet how sweet the air of the fields! Therefore, if by growing vegetables we can convert an expensive and objectionable material into a beneficial and profitable one, should we not be to blame were we not to take advantage of what we are offered by Nature? Next comes the land necessary for the poultry homes. In a sanitary point of view, these buildings ought to stand at least fifty feet apart, to allow for an efficient supply of fresh air, light, and sun; ought we to lose the advantage of rendering profitable land so conveniently situated, when three fourths of the poultry food should consist of green vegetables? The poultry homes must also be heated by hot-water pipes in winter; and why should we not render them serviceable in summer for irrigating or watering the land between the buildings? Lastly, why should the laborers of the poultry establishment not be profitably employed in their leisure time in attending on the adjacent land?

Vegetable Growing or Market Gardening,

I believe, is generally admitted to be highly profitable, and that a quick succession of crops can only be obtained from a plentiful supply of rich manure.

A market gardener is obliged to purchase his manure, to pay labor, rent, and taxes; he has no valuable use for his waste or weeds; the worms and slugs are destructive to his crops, and in dry weather he is either obliged to incur great expenses for watering, or sustain the destruction of his plants.

Poultry Breeding and Vegetable Growing.

Poultry breeding and vegetable growing ought, therefore, to be carried on conjointly, as the waste, weeds, inferior vegetables, worms, and slugs are valuable food for poultry; and the profit derived from choice vegetables ought to pay for the poultry's keep; under any circumstances this plan cannot be considered mere theory, as it must be obvious that where both branches are profitable separately, they must be still more so carried on conjointly and contiguously

Estimate of Revenue and Expenses

For the Year beginning July 1, 1865, and ending June 1, 1866.

EXPENSES.

	£	s.	d.
Cost of Buildings *	500	0	0
Cost of Plant and Materials	300	0	0
Cost of Stock as per annexed Details	125	0	0
Food and Working Expenses	889	10	0
Total	£1,814	10	0

* As at present projected, the building will be three hundred feet long, and contain fifty homes for fowls, and fifty for chickens; every home will consist of two rooms, one open and one closed, each twelve feet long and three feet wide.

POULTRY BREEDING IN

REVENUE.

	£	s.	d
Value of Buildings	500	0	0
Value of Plant and Materials	300	0	0
Value of Old Stock	125	0	0
Value of Young Stock	2,500	0	0
From Eggs	416	13	4
	£3,841	13	4
Deduct Expenses	1,814	10	0
Leaving a Net Profit of	£2,027	3	4

PROPOSED STOCK.

LAYING STOCK.

Cocks.	Hens.					
—	400	Common Fowls, at 3s. each	60	0	0

BREEDING STOCK.

Cocks	Hens					
1	6	Crève Cœur				
1	6	La Fleche				
1	6	Houdan				
1	3	Dorkings, gray				
1	3	" partridge				
1	3	Cochins, buff				
1	3	" partridge				
1	3	Bramahs, light				
1	3	" dark				
1	3	Spanish, black				
1	3	Hamburghs, golden				
1	3	" silver				
1	3	Poland, black				
1	3	" golden				
14	51	= 65 Birds, at 20s. each	65	0	0
		Total Cost of Stock	£125	0	0

Which will produce about 6120 Eggs for hatching, taken at an average of 120 Eggs for each Hen per annum, and allowing 1120 Eggs for casualties, will give 5000 Chickens.

As the Company's object is to sell eggs as well as rear poultry, the proportion of the above laying stock to the breeding stock will be found the most economical. Taking the average of eggs from a hen to be one hundred and twenty per annum, this gives about fifty thousand eggs from the laying stock, besides hatching about five thousand eggs from the breeding stock, allowing even only one sitting per annum to each hen; and as chickens begin to lay at six months old, and making a fair allowance for male birds, we may anticipate fifty thousand eggs from them within the next twelve months.

WORKING.

EXPENSES.

	£	s.	d.
Keep of 465 Old Birds, at 6s. per head	139	10	0
Keep of 5,000 Chickens till twelve months old, at 3s. per head	750	0	0
	889	10	0

REVENUE.

	£	s.	d.
100,000 Eggs, at 1d. each	416	13	4
5,000 Chickens, at 10s. each	2,500	0	0
	2,916	13	4
Deduct above Expenses	889	10	0
Leaving a Net Profit of	£2,027	10	0

Thus taking the cost price of the parent birds at twenty shillings each, and the young ones only at ten shillings each, we obtain in the first year a most valuable stock, part of which we can dispose of, as well as the stock of common fowls.

In the foregoing I have made no allusion to artificial hatching, which can be entirely dispensed with in an establishment where the object is not to rear poultry only, but also to produce eggs for the market; and you will perceive that the commercial success does not depend on artificial hatching, but on a judicious system of housing, feeding, and rearing poultry; yet we shall avail ourselves of artificial hatching at a period of the year when it is highly profitable to hatch chickens, and at a time when hens are not broody; but even at the most inclement season I anticipate that we shall have a great number of broody hens, on account of the genial temperature we shall be able to maintain in the building.

As regards the provisions made to rear chickens with less casualties than by the present system, I doubt not that they will prove as satisfactory as they are economical.

I remain, Gentlemen,
Your obedient Servant,
GEO. K. GEYELIN.

REPORT OF MR. GEYELIN

ON

THE POULTRY ESTABLISHMENTS IN FRANCE.

July 10, 1865.

Gentlemen:

Having at your request undertaken a journey to France with a view to promote the interest of our Company, I now beg to lay before you my observations on the subject, and which for more conciseness I have arranged under the following headings:—

1. The Object of the Voyage.
2. Natural and Artificial Incubation.
3. Rearing Poultry in France.
4. Fattening and Feeding.
5. Killing and Dressing.
6. Utilizing the Waste Products.
7. The System of Selling.
8. The Distinct Breeds.
9. Caponage and Virgin Cocks.
10. Opinions on my System of Poultry Breeding and Rural Economy.
11. Analysis of my Observations.

1. The Object of the Voyage.

The primary object of the voyage was to ascertain everything connected with poultry breeding in France, which might assist in promoting the success of our undertaking; also to inquire into the truth of numerous assertions in the public papers, that there existed in the vicinity of Paris most extensive Gallinocultural establishments, which by their particular system of artificial incubation, rearing, and feeding poultry on horseflesh, realized in one instance, viz., in that of M. de Soras, upwards of £40,000 per annum. I need scarcely say that, after the most searching investigation within a radius of forty miles of Paris, my opinion has been fully confirmed that such establishments do not nor can possibly exist; moreover, I can now firmly assert that there is not one establishment in existence within fifty miles of Paris where poultry breeding is carried on otherwise than on the old farm system; in fact, as you will perceive hereafter, I have spared neither time nor expense in this inquiry: yet, although I have been unable to trace anything like a system in poultry breeding in France at all approaching to that we are about to carry out, it cannot be denied that I have obtained very valuable information, which will, no doubt, prove of great advantage to our Company: moreover, I observed such startling novelties connected with poultry breeding in France, that I deemed it to the interest of our society that at least two of the directors should come there also to enable them to corroborate this report, which otherwise might have borne the appearance of exaggeration in many respects, and perhaps have still left a doubt in the mind of many persons whether I really made all possible inquiries into the truth of the reported existence of Gallinocultural establishments in France.

I will now briefly relate the steps I have taken to inquire into this matter. At the Jardin des Plantes of Paris, which corresponds to our Zoölogical Society in Regent's Park, also at the Acclimatation Society in the Bois de Boulogne, where the various breeds of poultry form an important object, the existence of any such Gallinocultural establishments in France was totally unknown; and they observed very justly that if any such really were to exist, they would be the first to know of it. Next I called three consecutive market days at the wholesale poultry market, La Vallée, Paris, where all the poultry, dead or alive, forwarded from the various parts of France, is sold by auction from five till nine o'clock in the morning. Several agents and poulterers made inquiries for me of poultry merchants from the different parts of France, but with the same result. I made further inquiries at the dead poultry market at the Halles Centrales, also of a number of fancy poultry dealers, but all to no purpose; a few days later, on calling again at the Jardin d'Acclimatation, Monsieur A. Geoffroy St. Hilaire, the director, told me that a friend of his had informed him that such an establishment really did exist at Mouy, near Beauvais in Picardie, and he gave me his card, and the following in writing, adding, however, that he did not believe in it, and that he should feel obliged by my communicating to him the result of my investigation : —

" They tell me that M. de Soras has at Mouy, near to B., a large poultry-breeding establishment; but if my inquiries are right he ought to have at Mouy 12,000 fowls, with which he supplies the Paris markets."

I then telegraphed the following: " De Soras, M. (express) B. Have you an establishment for poultry breeding? Reply by return of mail.
GEYELIN."

At the same time I posted a letter to the same effect, and asking permission to visit the establishment. The reply to the telegram was — *not known;* the letter as yet has not been returned; but to make the inquiry triply sure, I started myself for Mouy; arrived at Reil Junction, I was informed that such an establishment really did exist at Mouy, and within half a mile of the railway station, which news delighted me, to know that my journey was not like a wild-goose chase; therefore, on arriving at Mouy, I proceeded at once to the poultry establishment, but not of M. de Soras, whose name is not even known to any person in that neighborhood, but of M. Manoury, éleveur à Angy près Mouy, to whom I briefly related the object of my call. I was received with every courtesy, and informed that he knew of no such name as M. de Soras, nor of any establishment of the kind, but that he devoted his time to rearing some five thousand heads of poultry per annum; he neither fed them on horseflesh nor supplied the markets of Paris; that he sold none but pure breeds, and those to gentlemen and fancy poultry dealers; nevertheless, that his system of hatching, rearing, and feeding was so different to that adopted by others that it might possibly have given rise to those exaggerated reports; after which he conducted me over his establishment, and explained most minutely the system he has adopted, which, however, I need not explain in this part, as I shall have to refer to it under the several headings. I will now conclude by adding, that I have visited all those places in France so justly famed for their poultry, and from which those celebrated breeds of Houdan, La Fleche, and Crêve Cœur are obtained, where, also, I met with the utmost courtesy in my inquiries, though I had been informed that the farmers never explained or showed their system of poultry rearing to any one, which possibly may be true as regards their countrymen.

2. Natural and Artificial Incubation.

Of artificial incubation I have observed four different systems, which, although said to answer well, are yet far from being applicable to hatching in a commercial point of view. It matters, indeed, very little what system is adopted, provided the heat is maintained at an even temperature: to obtain this, various regulators have been invented, but none of which can as yet dispense with personal care. They all say that their regulators are perfect if the temperature of the room can be kept at the same degree of heat during incubation; that then they can regulate the heat of the incubator to any given degree; but as such conditions of a uniform temperature are impossible to maintain, considering the variations in the temperature of the atmosphere, I consider artificial hatching too expensive for ordinary purposes, and only to be adopted at certain times of the year, and then only in establishments where the heat can be maintained at a uniform temperature, day and night, by personal care.

At the Jardin des Plantes, in Paris, the manager of the poultry department, M. Vallée, employs an apparatus of his own invention, which he has patented, and for which he has obtained prizes at two exhibitions. The principle consists of water heated by means of a lamp as a medium for hatching: the temperature is regulated by admitting more or less cold air by means of a valve opened or closed by a mercury float.

At the Jardin d'Acclimatation two systems of artificial incubation are in use, and although both are on the hot-water principle, yet they differ materially: the one is heated by means of a lamp, and the temperature regulated by a valve admitting more or less cold air, and which is effected by a piston acted upon by the expansion or condensation of air under different temperatures; the other consists merely of a zinc box covered with non-conducting materials. This apparatus requires neither lamp, regulator, or thermometer; the hot water is renewed every twelve hours; and it is said to answer admirably. The eggs are placed in a drawer underneath the water tank, but I cannot help thinking that with an atmospheric temperature at or below freezing. point it would be very difficult to prevent the rapid cooling of the water.

The next and last system of artificial hatching I shall notice is that shown to me by M. Manoury at Mouy. It consists of an ordinary wine cask lined on the inside with plaster of Paris. In this cask several trays with eggs are suspended, and the top of the cask is provided with a certain number of vent-holes for admitting air, which is regulated by means of vent-pegs: the cask is surrounded to the top with a thickness of about four feet of horse manure. Though I am assured that this principle answers well, I entertain serious doubts about it for the same reasons as before stated.

The Natural Hatching

Differs entirely from what I ever saw before, and in some parts of France forms a special trade carried on by persons called *couveurs*, or hatchers. These hatch for farmers at all times of the year at so much per egg, or purchase the eggs in the market, and sell the chickens, as soon as hatched, from threepence to sixpence each, according to the season of the year. This system may aptly be called a living hatching machine, and, in my opinion, it is the very best and cheapest way of hatching, as will be seen by the following description: —

The Hatching-Room

Is kept dark, and at an even temperature in summer and winter. In this room a number of boxes, two feet long, one foot wide, and one foot six inches deep, are ranged along the walls. These boxes are covered in with lattice or wire work, and serve for turkeys to hatch any kind of eggs. Similar boxes, but of smaller dimensions, are provided for broody fowls. The bed of the boxes is formed of heather, straw, hay, or cocoa fibres; and the number of eggs for turkeys to hatch is two dozen, and one dozen for hens.

At any time of the year, turkeys, whether broody or not, are taught to hatch in the following manner: Some addled eggs are emptied, then filled with plaster of Paris, then placed into a nest; after which a turkey is fetched from the yard, and placed on the eggs, and covered over with lattice: for the first forty-eight hours she will endeavor to get out of her confinement, but soon becomes reconciled to it, when fresh eggs are substituted for the plaster of Paris ones; they will then continue to hatch, without intermission, from three to six months, and even longer; the chickens being withdrawn as soon as hatched, and fresh eggs substituted: after the third day the eggs are examined, and the clear eggs withdrawn, which are then sold in the market for new laid; but, as they may be soiled or discolored from having been sat upon, they clean them with water and silver-sand to restore their original whiteness.

The turkeys are taken off their nest once a day, to feed, and to remove their excrements from the nest; but, after a while, they cease self-feeding, when it is necessary to cram them, and give them some water once a day.

Amongst some places I visited, in company with two of your shareholders, may be mentioned the farm of Madame La Marquise de la Briffe, Chateau de Neuville, Gambais, near Houdan, where we observed twelve turkeys hatching at the same time; here, also, we witnessed the rearing and fattening, which will be alluded to hereafter. In another place,—that of Mr. Auché, of Gambais, a hatcher by trade,—we observed sixty turkeys hatching at the same time; and we were informed that, during winter and early spring, he had sometimes upwards of one hundred hatching at the same time, and that each turkey continued hatching for at least three months. At the farm of Mr. Louis Mary, at St. Julien de Faucon, near Lizieux, in Calvados, I saw a turkey that was then sitting, and had been so upwards of six months; and as I considered it rather cruel, the hatcher, to prove the contrary, took her off the nest, and put her in the meadow, and also removed the eggs; the turkey, however, to my surprise, returned immediately to her nest, and called in a most plaintive voice for her eggs; then some eggs were placed in a corner of the box, which she instantly drew under her with her beak, and seemed quite delighted. Moreover, I was informed that it was of great economical advantage to employ turkeys to hatch, as they eat very little, and get very fat in their state of confinement, and therefore fit for the market any day.

3. The Rearing of Poultry.

It seems strange that although in all countries the great difficulty of poultry breeding is the successful rearing, that no adequate means have ever been devised to counteract the influence of climates. In France, like here, a cold or wet spring is equivalent to a great loss in poultry, and it seems to be admitted everywhere that cold and wet do not agree with poultry; therefore, were it not for some novelties I observed in the rearing, to which I shall allude presently, I might well have said that their system is no better than our own; in fact, they show an utter disregard of all sanitary considerations; and without wishing to particularize any establishment, whether public or private, I may state that even the best conducted left room for great improvement in this respect. In some parts of France, where poultry breeding is carried on as a trade, they seldom allow a hen to lead the chickens after being hatched, as the hen is more valuable for laying eggs; but they intrust this office either to capons or turkeys, who are said to be far better protectors to the chickens than a hen: they require, however, a certain amount of schooling preparatory to being intrusted with their charge, and which consists in this: When a turkey has been hatching for some months, and shows a disposition to leave off, a glassful of wine is given her in the evening, and a number of chickens are substituted for the eggs; on waking in the morning, she kindly takes to them, and leads them about, strutting amidst a troop of seventy to one hundred chickens with the dignity of a drum-major. When, however, a troop leader is required that has not been hatching, such as a capon or a turkey, then it is usual to pluck some of their feathers from the breasts, and to give them a glass of wine, and, whilst in a state of inebriation, to place some chickens under them: on getting sober the next morning, they feel that some sudden change has come over them; and as the denuded part is kept warm by the chickens, they take also kindly to them.

Another important matter in rearing poultry is their

feeding, which differs also very much with our own, but which I shall have to notice under a subsequent heading.

In conclusion, I feel in justice bound to say that these artificial living protectors are most efficient to shelter chickens in the daytime, and in the evening they are placed with their charge in a shallow box filled with hay, from which they do not move till the door of the room is opened next morning. I must not omit to mention that the chickens are not intrusted to the mother, or a leader, before they are a week old, and then only in fine weather.

4. Feeding and Fattening.

The system of feeding poultry in France is far more judicious than our own; and I may safely assert that I have not noticed a single instance of poultry being fed on whole grain, as it is the case with us. On inquiring the reason why they fed by meal made into a stiff paste, I was informed that whole grain would be too expensive, produce less eggs, too much fat, and cause more disease when the fowls are fed *ad libitum*, so as to completely fill their crop, which renders the digestion difficult. The food is mostly composed of about one half bran and one half buckwheat, barley, or oatmeal made into a stiff paste, with which the fowls are fed twice a day, namely, at sunrise and sunset; this diet is given indiscriminately to old and young. In some farms, where the poultry have not the run of meadows, they are provided with a certain amount of animal and vegetable food, which system is so consonant with my own notion that I will now describe that followed at an establishment already noticed. All the waste of butchers' shops are obtained at the expense of collecting them; these are boiled, the fat skimmed off, which, when coagulated, is with the waste finely minced, and mixed with the meal; after which the waste of the kitchen garden, such as cabbage-stalks, are boiled in the liquid, and mixed with bran, sour poultry food, &c., which is then given to the pigs, who thrive admirably on it. Buckwheat is considered preferable to all other grains as a stimulant to laying eggs, and in winter a certain amount is given whole. The chickens, for the first week after being hatched, and in winter for a much longer time, are fed by hand on barley-meal mixed with milk, stale bread soaked in water, and green food finely chopped.

The Fattening of Poultry

Whilst the rearing is carried on by farmers, the fattening forms quite a special trade, and chiefly in the hands of cottagers, who purchase the chickens either from farmers or in the market; moreover, it is the exclusive trade of a few villages in each poultry breeding district, such as Goussainville, de Saint Lubin, de la Haye, &c., near Houdan, Villaine, and Boce, near La Flêche au Mans; also some hamlets near Saint Pierre Dive, Lizieux, Calvados. In these localities the system of fattening differs, however; the one consists of liquid cramming with barley-meal and milk, given by means of a funnel introduced into the throat of the fowl three times a day; this process is exceedingly expeditious, as one person can easily cram at the rate of sixty fowls per hour, and the fattening lasts from fourteen days to three weeks, according to the disposition of the chicken to take fat; the selection of the fattening stock requires some judgment, as some chickens are constitutionally too weak, and others have not the frame to receive fat. This system of liquid cramming is principally adopted in the neighborhood of Houdan; and to give an idea of the importance of this trade, I will now give a short extract from the pamphlet I was kindly presented with from a most intelligent agriculturist, Monsieur De la Fosse, Proprietaire à Orval, Goussainville près Houdan: —

"It is to be desired that our excellent and pure breed of Houdan should be propagated in every other country as much as it is in our own, where the poultry trade has taken such a development that it forms one of the principal sources of riches. A few exact statistics of this trade in our immediate neighborhood will give a correct idea of its importance. At the markets of Houdan, Dreux, and Nogent le Roi, there are sold annually upwards of six million heads of FAT poultry, namely: —

	Per Week.	Per Month.	Per Year.
Houdan	40,000	160,000	1,920,000
Dreux	50,000	200,000	2,400,000
Nogent le Roi	35,000	140,000	1,680,000
Total			6,000,000 "

This does not include the sale of chickens and poultry, which forms a separate trade.

Monsieur De la Fosse also deprecates the use of fat for fattening purposes, as it deteriorates the fineness and flavor of the flesh. In the districts of Le Mans and Normandy, the fattening is performed by dry cramming, viz.:

the meal of barley and buckwheat is made into a stiff paste with milk and water, then formed into pills two inches long and half an inch in diameter; these are dipped into water, and forced into the throat of the fowl, until the crop is filled, twice a day; it is, however, of importance not to cram a fowl until she has digested the previous meal, as otherwise it might produce inflammation and death.

A most ill-founded notion prevails with all fatteners — that poultry will fatten much quicker without light or ventilation, and without ever removing their excrements, which makes these places most offensive and unhealthy; no other reason could be assigned to me than that they were quite sure that the smell of the excrements stimulated the fattening; in this there is about as much reason as in the notion our farmers used to entertain that pigs could only thrive in filth. In one place, however, which I visited in company with Monsieur Noel, proprietor of the Lion d'Or at La Flêche, a most intelligent man, and himself a large farmer, the cottager had provision made for the excrements to fall through the floor of the pen; and on pointing out the innovation, he prided himself on his invention, as, said he, I can now remove the manure, and the feathers of the fowls get less dirty, and the birds have also more air. This, surely, is a step in the right direction.

5. Killing and Dressing.

This also is a speciality, carried on by men called *Tueurs et Apprêteurs;* they are astonishingly expert in their business; and unless witnessed, as we have done, it would appear incredible that one man can kill and pluck at the rate of one fowl per minute, or sixty per hour: the price paid for this work is about one farthing per head for lean and one halfpenny for fat poultry. The system of killing differs, however, in this, that whilst in Paris they make a gash in the throat, in the country they stick the poultry in the back of the roof of the beak; but both cause immediate death; the latter, however, is the cleanest and most desirable. They deprecate our system of twisting the neck, as cruel, discoloring the flesh, and causing early putrefaction of the coagulated blood. When a man kills, he has three baskets near him, into which he drops the feathers according to size; and the reason of plucking the fowls instantaneously after death is the great saving in time, and the prevention of tearing the skin, which latter cannot well be avoided when the fowl once gets cold.

The Dressing.

The lean fowls are immediately emptied of their intestines; but not so with the fat stock, which contain a large quantity of valuable fat, which is used for basting, and to give flavor to lean poultry.

With chickens they take care to leave the down on, as an index of their age, and in all fowls they leave about half a dozen feathers in the rump, which gives a very pretty appearance.

As soon as the fowl is plucked, and before cold, it is laid on its back on a bench, and wrapped round with a wet linen cloth to mould its shape, and to give the skin a finer appearance; however, they use no flour, as with us, to give an old hen the appearance of a chicken.

The fat poultry is drawn and dressed by cooks; they make an incision under the leg to withdraw the intestines, by which means the fowl is not disfigured.

6. Utilizing the Waste Products.

Poultry Manure.

In France, as well as in our own country, most eminent chemists have proved by analysis that poultry manure is a most valuable fertilizer; and yet, for want of a proper system in housing poultry, it has as yet not been rendered available to rural economy. The celebrated Vauquelin says that when the value of manures is considered in relation to the amount of azote they contain, the poultry manure is one of the most active stimulants; and when, as a means of comparison, the following manures are taken in parts of 1000, it will be found that,—

Horse Manure contains	4.0	parts of azote.
Guano, as imported,	49.7	" "
Guano, when sifted of vegetables and stones,	53.9	" "
Poultry Manure,	83.0	" "

In France, as in England, the poultry manure is left to accumulate in the poultry homes, to the loss of farmers and to the detriment of the health of fowls.

The Feathers

Are carefully collected and sorted, and when well dried sold to dealers.

The Intestines

Are boiled, the fat skimmed off, which is sold separate; the intestines are then minced as food for poultry, and the liquid is used for feeding pigs.

The Combs and Kidneys

Are sold to pastry-cooks — the first for decorating and the latter for flavoring pies.

The Head, Neck, and Feet

Are sold to hotels, restaurants, &c., for flavoring sauces, or boiled down to make chicken jelly.

7. THE SYSTEM OF SELLING

Poultry in France is far preferable to our own, although, in my opinion, it would be still better were poultry sold by weight. However, a farmer or merchant who consigns poultry to Paris is sure to obtain a true return of whatever they fetched, as he does not rely, as with us, on the honesty of a dealer.

At the wholesale poultry market, La Vallée, in Paris, where all poultry, dead or alive, is forwarded from all parts of France, there are a number of licensed agents to whom the poultry is consigned, and who sell it by auction to the highest bidder; this market is a curious scene, and worth seeing, from four till nine in the morning, where thousands of crates, of all descriptions of poultry, are disposed of, and cleared out, before twelve o'clock in the day.

Every village has its weekly markets, where farmers and their wives bring their produce for sale, in preference to selling it at the farm-yard. The police regulations in these markets are strictly enforced. The various products are classified before the market begins. Each person is bound to keep his assigned place, and not allowed even to uncover his goods, and much less to sell, before the bell rings, under a fine of five francs. At the ringing of the bell, the bustle to uncover, the rush of buyers, and the chattering, are worth while to witness. The dealers and merchants take up their stand outside the market, where they send all the products they purchase. The seller has a ticket given him, with the purchase price on it, and is paid on delivery of the goods at the dealer's stand. It seems almost incredible to believe, that even in some village markets, within two hours, such a vast amount of business can be transacted with the greatest order and decorum. Some merchants will purchase from two thousand to three thousand pounds of butter; others, twenty thousand to thirty thousand eggs, or some thousand heads of poultry, &c.; all which are taken to their warehouse to be sorted, packed, and perhaps forwarded the same day either to London or Paris. I may add, that the current price for every commodity is fixed and known immediately after the market opens, and depends entirely on the demand and supply. For instance, fat chickens fetched four shillings each; twelve eggs, sevenpence; butter, tenpence per pound, &c.

For the foregoing information, I am mainly indebted to Mr. H. Lindon, Jr., a most obliging gentleman, who represents at Lizieux the Messrs. Lindon Brothers & Co. of

London, general merchants; in his company I have visited several farms, and attended market, at which he makes purchases of butter, grain, &c., for the London market.

8. The Distinct Breeds.

There are three perfectly distinct breeds, all very characteristic in their appearance; and, when of pure race, they are very true to all their points. I shall only give a cursory description of the appearance of those we have now at our establishment in Bromley.

Houdan Fowl.

Whatever has been said to the contrary, this breed, when pure, is most characteristic; but it must be admitted, that most of the farmers near Houdan know as little of the pure Houdan breed as those of La Flêche and Crêvecœur know of theirs; and, if you were to order some first class birds of them, irrespective of price, they would with good conscience forward fowls of a large size — but, from a want of knowledge, some cross breeds. To illustrate this, I may mention that I could have purchased, at the markets in those respective localities, splendid thorough-bred specimens for about three shillings, the price of common fowls, — but which were worth in France even one pound each. There are, however, in each locality, some persons who take an interest in their pure breeds, particularly since they have been encouraged by the award of prizes from poultry exhibitions.

The Houdan fowl has a very bulky appearance, its plumage invariably black and white spangled — a crest of the same color; comb, triple, the outsides opening like two leaves of a book, and the centre having the appearance of an ill-shaped long strawberry. With the cock the comb is very large, whilst with the hen it ought to be scarcely perceptible. The legs are strong, and of a lead color, with five claws, the two hind ones one above the other. Strongly-developed whiskers and beards both in cocks and hens. This is one of the finest races of fowls, but its qualities surpass even its beauty; besides the smallness of their bones, the fineness of their flesh, they are of an extraordinary precocity and fecundity; they lay large and white eggs, and the chickens are fit for the table at four months old. It is, however, observed that they are very indifferent for hatching. The weight of adults is from seven to eight pounds, in which the bones figure for one eighth. The chicken, when four months old, weighs, without the intestines, about four and a half pounds.

The Crêvecœur Race,

In outward appearance, resembles closely the Houdan, but its crest, whiskers, and beard are still more developed; the crest is only double, and projects like two horns with the cock, but with the hen it is very small; the whole plumage ought to be perfectly black, although there are some white, also blue varieties, which are, however, only a degeneration; legs black; the claws, four in number, are stronger and longer than those of the Houdan. This breed is said to be superior, in all respects, even to

Houdan fowls, and justly esteemed as the most precocious and finest in the world, as the chickens are fit for the table at three months old, and at six months old weigh from seven to eight pounds when fattened; the eggs also are larger, and of a beautiful white.

La Flêche Race.

This breed differs entirely with the two preceding ones. In appearance they resemble the Spanish; the plumage, which is jet black, fits close to the body, and gives an idea of less bulk than those of Houdan and Crêvecœur, although they actually are heavier fowls. They are very long in coming to maturity, but which happens generally at the season when poultry is most scarce, on account of which, coupled with the exquisite flavor of the flesh, they fetch fabulous prices; and even at the time I was at La Flêche, the beginning of July, the fat chickens sold in the market at five shillings each. Skin white, fine, transparent, and very elastic, which enables them to take an extraordinary amount of fat. The weight of adults is from eight to ten pounds, and the bones less than one eighth of the weight; when standing erect they measure twenty-two inches in height and twenty-three inches in circumference, taken from under the wings. The legs, and four claws, of a lead color, are strong; the comb in appearance like two horns, like those of Crêvecœurs, with a little crest behind; the face white, like the Spanish, and a horn on the beak like that of the rhinoceros, form the principal characteristics of this fine race. It is said they begin to lay early in the year; but their eggs, though abundant, are smaller than those of other

French breeds, and as regards sitters they are considered as bad as the Spanish.

9. Caponage and Virgin Cocks.

There seems to exist a considerable difference of opinion in various parts of France as to the necessity of castrating young cocks for fattening purposes. In some localities they pretend that when cocks are not allowed to associate with the opposite sex, they will attain, when fattened, a greater weight, and be much finer as regards flavor of flesh; others again say that when a cock is castrated, it can be kept till a more mature age without deteriorating its quality, and by this attain an extraordinary weight when fattened, besides making them useful as troop leaders of chickens, as before described. I cannot decide which of the two systems is the best or most advantageous, any more than I can decide about the two systems of cramming, without making experiments; this much, however, I have noticed, that virgin cocks fatten very readily, and fetch prices as high as capons.

10. Opinions on my System of Poultry Breeding and Rural Economy

Wherever I went and whenever I had an opportunity, I promulgated my system of poultry breeding in connection with rural economy with a view to elicit opinions, having been taught by experience that diversity of opinion is the greatest stimulant to improvement and progress. Without naming any individual opinion, I may state that, without exception, all were favorable as regards the practicability of the undertaking when carried out on an extensive scale, as then the working expenses would be at their minimum and the returns at the maximum; that they do not consider it difficult to rear chickens in-doors, as their winter and spring chickens are all reared in outhouses. Some, however, hold it to be beneficial for fowls to get wet, with which I differ, as they are not amphibious, and require only dry dust to clean themselves. The separation system is much approved of, as it enables the races to be kept pure, in which they find the greatest difficulties in farm-yards: the arrangement for nests, feeding, warming, and ventilation are likewise commended — in fact I was told several times, "Ah, you Englishmen, when you do anything you do it well and on a grand scale."

11. Analysis of my Observations.

Fiction, when well told and supported by imaginary statistics, bears often more semblance to truth than reality itself; this fact was never better illustrated than by the interesting account given by some ingenious and inventive mind of certain Gallinocultural establishments, whose illusive existence was stated to be in the vicinity of Paris, and where the exclusive diet of the fowls was horseflesh. The story seemed so plausible, and the details so minute, that it was accepted as a fact, and in due course published in numerous scientific and other papers of this and other countries; indeed, the fact that fowls are omnivorous, and that they have a predilection for animal food, is so well known, that had it not been explicitly stated that their exclusive diet was horseflesh, I should have credited it myself; my doubts did, however, not arise on account of the use of horseflesh, — which is just as good, and perhaps better, than many other animals' flesh for the food of poultry, — but solely on account of its pretended exclusive use. I have been informed at the Jardin des Plantes and at the Jardin d'Acclimatation, in Paris, that this subject has created as much interest and deception in other countries as our own, as persons from Russia, America, and other parts of the world, had come on purpose to Paris to visit those imaginary establishments. Whether on account of the daily increasing price of animal food the public mind was prepared to believe in the existence of such Gallinocultural establishments, where they slaughter fifty horses per diem for the food of poultry, or whether the publication of such fictions does more harm than good, I will not venture to discuss; nor can I say whether the persons who were disappointed in the object of their journey were compensated by learning some profitable matters not included in their programme of inquiry; but what I can assert, and which I believe will be fully borne out by the preceding report, is, that my journey to France will prove in many respects most beneficial to the interest of our Company. In support of this assertion I cannot do better than quote what I stated in my first report, viz., —

"We are not about to carry out any new invention in poultry breeding, but merely a wise combination of well-established facts: individually, the facts are well known; but a combination of them applied to poultry breeding has hitherto escaped the notice of rural economists."

Such is, in fact, the case in the undertaking we are about to carry out; namely, a combination and adoption of all the most successful systems in poultry breeding, whether of this or any other country; and it must be as satisfactory to you to know, — after having honored me with your confidence, — as it was pleasing to me to see, that the system of poultry breeding we are carrying out at Bromley, in Kent, combines every element of success.

With the exception of hatching by the aid of turkeys, rearing by turkeys and capons, and some other novelties connected with poultry breeding, and which we shall adopt, the soundness of my system is now proved by the successful working of its several parts in various places of France.

I can now with every confidence congratulate you for having founded the first Gallinocultural establishment in the world, and one that will prove as beneficial to you as it will be a boon to the nation at large; and which soon must become the national nursery for all pure races of poultry from whence farmers and others can be supplied with first class breeding stocks at moderate prices; an establishment, it is to be trusted, that will not only prove the foundation to an increase of animal food and the amelioration of poultry breeds, but also prove the means of poultry becoming cheaper and of more general use than it now is.

www.ingramcontent.com/pod-product-compliance
Lightning Source LLC
Chambersburg PA
CBHW082334220526
45470CB00008B/2506